VAUXHALL
NOVA
1983 to February 1992

All Vauxhall Nova models except Van and Diesel engine models. Does **not** cover revised Nova range introduced in February 1992.

(1820 – IWI)

A Haynes Handbook and Drivers Guide
© Haynes Publishing 1994

Printed and published by
J H Haynes & Co Ltd Sparkford
Nr Yeovil Somerset BA22 7JJ
England

ISBN 1 85010 820 X

British Library Cataloguing in Publication Data. A catalogue record for this book is available from the British Library

All rights reserved. No part of this book may be reproduced or transmitted in any form or by any means, electronic or mechanical, including photocopying, recording or by any information storage or retrieval system, without permission in writing from the copyright holder

HANDBOOK & DRIVERS GUIDE
by A K Legg LAE MIMI

Full details of all servicing and repair tasks for the models covered by this Handbook can be found in the Owners Workshop Manual – OWM **909** Nova 1983 to February 1992.

ACKNOWLEDGEMENTS

We gratefully acknowledge the assistance of RoSPA in compiling the information used in this Handbook. Thanks are due to Duckhams Oils who provided lubrication data. Certain other illustrations are the copyright of Vauxhall Motors Ltd., and are used with their permission. Additional photographs were supplied by Quadrant Picture Library/Auto Express. Thanks are also due to all those people at Sparkford who helped in the production of this Handbook.

We take great pride in the accuracy of information given in this Handbook, but vehicle manufacturers make alterations and design changes during the production run of a particular vehicle of which they do not inform us. No liability can be accepted by the authors or publishers for loss, damage or injury caused by any errors in, or omissions from, the information given.

CONTENTS

About this Handbook	5
The Nova family	7
Buying and selling	9
General tips on buying	9
Points to look for when buying	10
General tips on selling	11
Dimensions and weights	12
Controls and equipment	15
Driver's instruments and controls	18
Interior equipment	27
Exterior equipment	34
Accidents and emergencies	35
How to cope with an accident	35
First aid	36
Requirements of the law	38
Essential details to record	38
Accident report form	39
How to cope with a fire	42
How to cope with a broken windscreen	42
What to do if your car is broken into	42
Breakdowns	43
Breakdowns on an ordinary road	43
Motorway breakdowns	44
Changing a wheel	44
Towing	46
Starting a car with a flat battery	47
What to carry in case of a breakdown	49
Driving safety	51
Before starting a journey	51
Driving in bad weather	52
Motorway driving	53
Towing a trailer or caravan	54
Alcohol and driving	56
Skid control	56
Advice to women drivers	57
Child safety	59
Driving abroad	61
Reducing the cost of motoring	65

Car crime prevention	67
Service specifications	69
Regular checks	73
Checking oil level	74
Checking coolant level	75
Checking brake fluid level	76
Checking tyres	76
Checking washer fluid level	78
Checking battery electrolyte level	78
Checking wipers and washers	79
Checking lights and horn	79
Checking for fluid leaks	80
Servicing	81
Service schedule	82
Safety first!	86
Buying spare parts	89
Service tasks	93
Seasonal servicing	107
Tools	109
What to buy	109
Care of your tools	109
Bodywork and interior care	111
Cleaning the interior	111
Cleaning the bodywork	111
Dealing with scratches	112
Bulb, fuse and relay renewal	113
Bulbs – exterior lights	114
Bulbs – interior lights	117
Fuses	118
Relays	119
Preparing for the MOT test	121
Fault finding	123
Car jargon	129
Local radio frequencies	139
Conversion factors	144
Distance tables	146
Index	148
Servicing notes	152

VAUXHALL NOVA

4 ABOUT THIS HANDBOOK

▲ *Nova 1.6 GSi Hatchback (1991)*

VAUXHALL NOVA

ABOUT THIS HANDBOOK

The idea behind this Handbook is to help you to get the most out of your motoring.

Apart from the things that every owner needs to know, to deal with unexpected mishaps like a puncture or a blown light bulb, you'll find clearly-presented information on road safety, hints on driving abroad, and tips on how to prepare your car for the MOT test. We've also included details of local radio frequencies to help you avoid those inevitable 'jams', and there's a Section on what to do in the unfortunate event of an accident.

For those not familiar with their car, there's a Section to explain the location and the operation of the various controls and instruments. Additionally, there's a Section on fault finding, and a useful glossary of car jargon.

Garage labour charges usually form the major part of any car servicing bill, and we hope to help you to reduce those bills by carrying out the more straightforward routine servicing jobs yourself. If you're about to start carrying out your own servicing for the first time, we aim to provide you with easy-to-follow instructions, enabling you to carry out the simpler tasks which perhaps you've left to a garage or a 'car-minded' friend in the past. Even if you prefer to leave regular servicing to a suitably-qualified expert, by using this book you'll be able to carry out regular checks on your car, to make sure that your motoring is safe and hopefully trouble-free. You'll also find advice on buying suitable tools, and safety in the home workshop.

Some readers of this Handbook may not yet have bought a Vauxhall Nova, so we've included a brief history of the range, and some useful tips on buying and selling.

All in all, we hope that this book will prove a handy companion for your motoring adventures, and hopefully we'll help to reduce the problems which inevitably crop up in everyday driving. If you're bitten by the DIY bug, and you're keen to tackle some of the more advanced repair jobs on your car, then you'll need our **Owners Workshop Manual** (OWM 909). This manual gives a step-by-step guide to all the repair and overhaul tasks, with plenty of illustrations to make things even clearer.

Happy motoring!

THE NOVA FAMILY

▲ *Nova Merit 1.4i Cat Hatchback (1990)*

▲ *Nova 1.2 L Saloon (1983)*

THE NOVA FAMILY

The Vauxhall Nova was launched in the UK in early 1983, and was the General Motors UK version of the Opel Corsa, which had already been established as a popular small car on the Continent. The launch was not without controversy, as the factory unions were originally against the Nova's importation from GM's factory in Zaragoza in Spain. This dispute was eventually resolved, and the Nova has become a very popular car in the UK.

Initially, the Nova was available as a three-door Hatchback or a booted two-door Saloon, with a choice of engine size from the overhead valve 1.0 litre to the 1.2 & 1.3 litre overhead camshaft engines. Four-door Saloons and five-door Hatchbacks were added to the range in 1985.

The 1.4 litre overhead camshaft carburettor engine was introduced in late 1989, then in August 1990 the same 1.4 litre engine was given single point fuel injection together with a catalytic converter. The largest engine in the Nova range, the 1.6 litre overhead camshaft multi-point fuel injection engine, was fitted to the GTE which was introduced in May 1988. The GTE is fitted with a special close-ratio gearbox, and sprints from 0 to 60 mph in less than 9.5 seconds, with a top speed of around 116 mph. The GTE model was redesignated the GSi in November 1990, at which time the whole range was extensively facelifted.

All engines are of 4-cylinder in-line type, mounted transversely at the front of the car, with the gearbox positioned on the left-hand end of the engine (when viewed from the driver's seat). From the start of production, both four and five-speed gearboxes were available according to model.

The Nova's independent suspension is of well-proven standard type, consisting of MacPherson struts and coil springs at the front, and trailing arms and transverse torsion member with coil springs at the rear. Front and rear anti-roll bars are fitted according to model.

Novas are available with a variety of trim levels ranging from the basic L, through the SR sports version to the GL luxury version. All SR and GL models are fitted with a comprehensive array of instruments and driver comforts, and include central locking, a stereo system and tinted glass.

Most of the mechanical components of the cars are fairly conventional, and the main systems are designed to keep servicing time and costs as low as possible. The electronic ignition system (fitted to later models), clutch, brakes, wheel bearings, etc are all routine-maintenance-free, and should only require attention when certain items need to be renewed.

Such is the popularity of these cars that all spare parts are readily available from a number of sources, and for those wishing to carry out their own servicing to keep running costs to a minimum, the Nova is one of the best small cars you can buy.

8 THE NOVA FAMILY

▲ *Nova Merit 1.2 Saloon (1991)*

▲ *Nova 1.0 5-door Hatchback (1986)*

VAUXHALL NOVA

BUYING AND SELLING

GENERAL TIPS ON BUYING
- **Don't rush out and buy the first car to catch your attention**
- **Always buy from a recognised dealer in preference to buying privately**
- **Have the car checked over by someone knowledgeable before you buy it**
- **In general, you get what you pay for!**

Before buying a second-hand car, it's worthwhile doing some homework to try and avoid some of the pitfalls waiting for the unwary. First of all, don't rush out and buy the first car to catch your attention (all that glitters is not gold!), and remember that much of the responsibility is yours when it comes to the soundness of the deal, especially when buying a car privately.

Wherever possible, buy from a recognised dealer, and check that the dealer is a member of the Retail Motor Industry Federation, as this will provide you with certain legal safeguards if you have any problems. If you buy from a dealer, you are covered by the Sale Of Goods Act, which in summary states that the goods must be fit for their intended purpose, the goods must be of proper quality, and the goods must be as described by the seller. If you're buying a car privately, ask to see the service receipts and the MOT certificates going back as long as the car has been in the possession of the current owner (this will help to establish that the car hasn't been stolen, and that the recorded mileage is genuine), and always ask to view the car at the seller's private address (to make sure that the car isn't being sold by an unscrupulous dealer posing as a private seller). Check the vehicle documents for obvious signs of forgery and, if in doubt, contact the DVLA and give them details of the Registration Document, as they will be able to run a check on its authenticity.

As far as the soundness of the car itself is concerned, a genuine service history is helpful. This is provided by the service book supplied with the car when new, which should be completed and officially stamped by an authorised garage after each service. Cars with a full service history (fsh) usually command a higher price than those without.

To check the condition of a car, a professional examination is well worthwhile, if you can afford it, and organisations such as the AA and RAC will be able to provide such a

BUYING AND SELLING

service. Otherwise, you must trust your own judgement, and/or that of a knowledgeable friend. Although it's tempting, try not to overlook the mechanical soundness of the car in favour of the overall appearance. It's relatively easy to clean and polish a car every week, but when were the brakes and tyres last checked? Above all, safety must always take priority. Don't view a car in the wet, as water on the bodywork can give a misleading impression of the condition of the paintwork. First of all, check around the outside of the car for rust, and for obvious signs of new or mis-matched paintwork which might show that the car has been involved in an accident. Check the tyres for signs of unusual wear or damage, and check that the car 'sits' evenly on its suspension, with all four corners at a similar height. Open the bonnet and check for any obvious signs of fluid leakage (oil, water, brake fluid), then start the engine and listen for any unusual noises – some background noise is to be expected on older cars, but there should be no sinister rattles or bangs! Also listen to the exhaust to make sure that it isn't 'blowing' indicating the need for renewal, and check for signs of excessive exhaust smoke. Black smoke may be caused by poorly adjusted fuel mixture, which can usually be rectified fairly easily, but blue smoke usually indicates worn engine components, which may prove expensive to repair. Finally, drive the car, and test the brakes, steering and gearbox. Make sure that the car doesn't pull to one side, and check that the steering feels positive and that the gears can be selected satisfactorily without undue harshness or noise. Listen for any unusual noises or vibration, and keep an eye on the instruments and warning lights to make sure that they are working and indicating correctly.

If all proves satisfactory, try to negotiate a suitable deal, but remember that a dealer has to work to a profit margin, and it's unlikely that you'll find a good car for a silly price. Always obtain a receipt for your money.

On the whole, it's true to say that you get what you pay for. Above all, don't be rushed into making a hasty decision.

POINTS TO LOOK FOR WHEN BUYING A VAUXHALL NOVA

In addition to the points outlined previously, there a few special points to look out for when buying a Nova.

The Nova is a sound car with above-average reliability, and has very low running costs compared with other cars of the same size. All the mechanical parts of the Nova have been tried and tested in the Vauxhall Astra, which is a very similar car, so all of the usual teething problems associated with a new car have been sorted out.

There are two types of engine – the overhead valve 1.0 litre engine, and the overhead camshaft 1.2, 1.3, 1.4 & 1.6 litre engines.

The 1.0 litre engine is very reliable and economical, and should cover up to 100 000 miles without any major expense. If the mileage is on the high side, it is worth checking the engine for excessive oil consumption. Warm up the engine, then leave it running while you open the bonnet and listen for unusual noises – it should idle smoothly without any misfiring, and there should be no excessive noise. If there is a clattering noise from the front of the engine, the timing chain may be worn and need replacing, and if there is a clattering noise from the top of the engine, the valve clearances may need to be adjusted. None of these jobs is expensive to put right, but it would be worth trying to negotiate a lower price to cover them. While the engine is ticking over, get someone to rev it up while you move to the rear of the car and check for blue smoke from the exhaust. A little black smoke is acceptable, but if there is even the slightest hint of blue smoke, the car should be rejected, as it is likely there will be some fairly expensive repairs needed.

The overhead camshaft 1.2, 1.3, 1.4 & 1.6 litre engines should also be checked for blue smoke as previously described, but in addition check for excessive noise from the camshaft (top of the engine), especially when the engine is revved up. Fitting a new camshaft is quite an expensive job, and allowance for this should be made when negotiating the price. However, once the job has been done, the

BUYING AND SELLING

engine should be good for another 50 000 miles or so, provided that the rest of the engine is in good condition. If the engine has covered more than say 40 000 to 50 000 miles, ask if the timing belt has been renewed – if you buy the car, it is worth fitting a new timing belt, to prevent expensive repairs which would be necessary if the belt was to break. On the test run, check for a flat spot or lack of response when accelerating quickly from a low speed. This may be due to faulty spark plug leads or a worn carburettor, although there could be other reasons. A new set of spark plug leads is not expensive, but if these have already been renewed, a new carburettor may be necessary, in which case try to negotiate a reduction to cover the extra expense.

The bodywork on all Novas is generally very good, and is not prone to bad rusting. The areas most likely to show a little rust are along the sills and bottoms of the doors, and on Hatchback models, along the bottom of the rear tailgate. Any accident damage which has remained unrepaired may also show signs of rusting, and if this is evident, the car should be rejected. Damage to the plastic bumpers can also be expensive. Unless abused, the upholstery is usually hard-wearing, and should not show any signs of deterioration. On Hatchback models, it is common for the rear seats to rattle and squeak a little on rough roads when not carrying rear passengers, and as there is not a lot that can be done to stop this, you will have to decide whether you can put up with it.

If you are looking for one of the sporty versions of the Nova, there are no specific problems to look out for, but do consider the depreciation factor. Particularly on late cars, values can drop dramatically, making the relative running costs appear high.

If you intend doing a lot of motorway driving, the 5-speed models would be preferable to the 4-speed ones.

Secondhand Novas are in quite plentiful supply, and you should not have any trouble finding a good example. Vauxhall new car prices have recently been reduced to encourage sales, and this has effectively pushed secondhand prices down, so now a secondhand Nova represents really good value.

GENERAL TIPS ON SELLING

- **Make sure that the car is clean and tidy**
- **Make sure that all of the service documents, registration document etc, are available for inspection**
- **Ask yourself ... 'Would I buy this car?'**

Obviously when selling a car, bear in mind the points which the prospective buyer should be looking for, as described in the above sections. It goes without saying that the car should be clean and tidy, as first impressions are important. Any fluid leaks should be cured, and there's no point in trying to disguise any major bodywork or mechanical problems.

If you're trading the car in with a dealer, you will always get a lower price than if you sell privately, but you can be fairly sure that there will be less comeback to you should any unexpected problems develop. If selling privately, don't allow the buyer to take the car away until you have his/her money, and it's a good idea to ask him/her to sign a piece of paper to say that he/she is happy to buy the car as viewed, just in case any problems develop later on. Give a receipt for the money paid.

12 DIMENSIONS AND WEIGHTS

Note: All figures are approximate, and will vary depending on model

DIMENSIONS [mm (in)]
Overall length

Hatchback (up to end October 1990)	**3622** (142.6)
Hatchback (from November 1990)	**3652** (143.8)
Saloon (up to end October 1990)	**3955** (155.7)

VAUXHALL NOVA

DIMENSIONS AND WEIGHTS

Note: All figures are approximate, and will vary depending on model

Dimensions continued [mm (in)]
Overall width

Hatchback (up to end October 1990)	**1532** (60.3)
Hatchback (from November 1990)	**1535** (60.4)
Saloon (up to end October 1990)	**1540** (60.6)
Saloon (from November 1990)	**1542** (60.7)

Overall height

Hatchback	**1365** (53.7)
Saloon	**1360** (53.5)

WEIGHTS [kg (lb)]
Nominal kerb weight

1.0 litre Hatchback	**735** to **785** (1621 to 1731)
1.2 litre Hatchback	**740** to **807** (1632 to 1779)
1.3 litre Hatchback	**750** to **825** (1654 to 1819)
1.4 litre Hatchback	**780** to **827** (1720 to 1824)
1.6 litre Hatchback	**818** to **865** (1804 to 1907)
1.0 litre Saloon	**740** to **790** (1632 to 1742)
1.2 litre Saloon	**740** to **809** (1632 to 1784)
1.3 litre Saloon	**750** to **820** (1654 to 1808)
1.4 litre Saloon	**805** to **824** (1775 to 1817)

Maximum roof rack load

2/3-door models (up to end October 1989)	**80** (176)
4/5-door models (up to end October 1989)	**50** (110)
All models from November 1989	**80** (176)

Maximum towing weight (for trailer or caravan)

1.0 litre	**500** (1103)
1.2 litre	**600** (1323)
1.3, 1.4 & 1.6 litre	**800** (1764)

Maximum trailer/caravan noseweight

All models	**50** (110)

VAUXHALL NOVA

14 CONTROLS AND EQUIPMENT

VAUXHALL NOVA

CONTROLS AND EQUIPMENT

For those not familiar with the Vauxhall Nova models, this Section will help to identify the instruments and controls. Typical instrument panel layouts are shown in the accompanying illustrations. The operation of most equipment is self-explanatory, but some items require further explanation to ensure that their use is fully understood.

Note that not all items are fitted to all models.

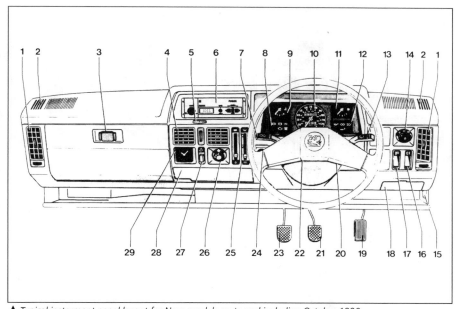

▲ Typical instrument panel layout for Nova models up to and including October 1990

1 Side air vents
2 Door window demister air vents
3 Glovebox
4 Centre air vents
5 Front/rear loudspeaker fader control
6 Radio, or cover plate with compartment for coins
7 Direction indicator switch
8 Warning lights
9 Engine coolant temperature gauge
10 Speedometer with mileometer and tripmeter
11 Fuel gauge
12 Warning lights
13 Windscreen/headlight wash/wipe switch
14 Lighting switch
15 Bonnet release lever
16 Rear foglight switch
17 Tailgate wash/wipe switch
18 Fusebox
19 Accelerator pedal
20 Ignition switch and steering column lock (not visible)
21 Brake pedal
22 Horn push
23 Clutch pedal
24 Choke
25 Heater air temperature and distribution controls
26 Heater fan motor/heated rear window switch
27 Hazard warning flasher switch
28 Ashtray with cigarette lighter
29 Clock

VAUXHALL NOVA

CONTROLS AND EQUIPMENT

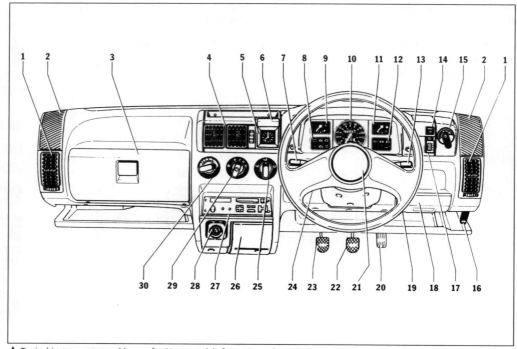

▲ *Typical instrument panel layout for Nova models from November 1990 on*

1 Side air vents
2 Door window demister air vents
3 Glovebox
4 Centre air vents
5 Clock
6 Hazard warning flasher switch
7 Direction indicator switch
8 Warning lights
9 Engine coolant temperature gauge
10 Speedometer with mileometer and tripmeter
11 Fuel gauge
12 Warning lights
13 Windscreen/headlight wash/wipe switch
14 Rear foglight switch
15 Lighting switch
16 Bonnet release lever
17 Headlight range adjustment
18 Fusebox
19 Ignition switch and steering column lock (not visible)
20 Accelerator pedal
21 Horn push
22 Brake pedal
23 Clutch pedal
24 Choke (not visible)
25 Air distribution control
26 Ashtray
27 Radio
28 Cigarette lighter
29 Heater fan motor/heated rear window switch
30 Heater temperature control

VAUXHALL NOVA

CONTROLS AND EQUIPMENT   17

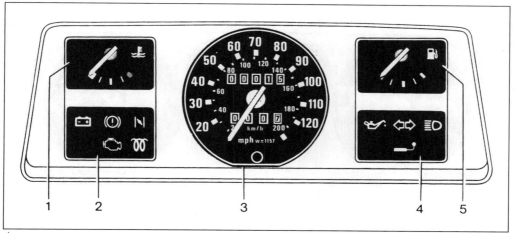

▲ Instrument panel arrangement on low-series models

1 Engine coolant temperature gauge
2 Warning lights
3 Speedometer with mileometer and tripmeter
4 Warning lights
5 Fuel gauge

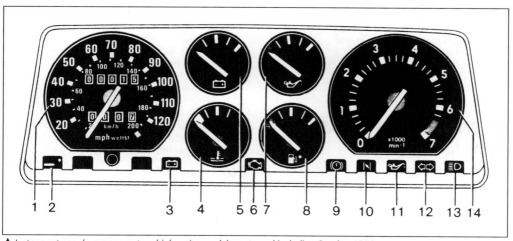

▲ Instrument panel arrangement on high-series models up to and including October 1990

1 Speedometer with mileometer and tripmeter
2 Warning light for trailer direction indicators
3 Ignition warning light
4 Engine coolant temperature gauge
5 Voltmeter or ignition warning light
6 Warning light for electronic engine management system
7 Engine oil pressure gauge, or warning light for engine oil pressure
8 Fuel gauge
9 Handbrake-on warning light
10 Choke warning light
11 Low oil pressure warning light
12 Direction indicator warning light
13 Headlight main beam warning light
14 Tachometer ('rev. counter')

VAUXHALL NOVA

CONTROLS AND EQUIPMENT

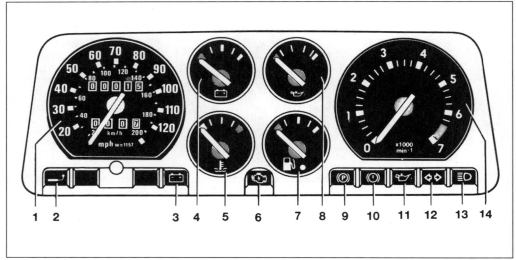

▲ *Instrument panel arrangement on high-series models from November 1990 on*

1   *Speedometer with mileometer and tripmeter*
2   *Warning light for trailer direction indicators*
3   *Ignition warning light*
4   *Voltmeter or ignition warning light*
5   *Engine coolant temperature gauge*
6   *Warning light for electronic engine management system*
7   *Fuel gauge*
8   *Engine oil pressure gauge, or warning light for engine oil pressure*
9   *Handbrake-on warning light*
10   *Brake hydraulic fluid level warning light*
11   *Low oil pressure warning light*
12   *Direction indicator warning light*
13   *Headlight main beam warning light*
14   *Tachometer ('rev. counter')*

DRIVER'S INSTRUMENTS AND CONTROLS

Speedometer

Indicates the car's road speed; incorporates a mileage recorder, and a tripmeter which can be reset by pressing the button protruding from the instrument.

Tachometer ('rev. counter')

Indicates engine speed, in revolutions per minute (rpm or rev/min). On early models, the tachometer scale is in multiples of 100, but on later models it is in multiples of 1000. For normal driving and the best fuel economy the engine speed should be kept in the 2000 to 4000 rpm (black) range. The red outlined zone indicates engine speeds which should only be used for brief periods. The red zone indicates the danger zone, and the engine may incur damage if the indicator enters this section.

Fuel gauge

Indicates the quantity of fuel remaining in the tank. When the needle enters the red band, the car should be refuelled.

Temperature gauge

Indicates the engine coolant temperature. As the engine warms up, the needle should move from the blue (cold) scale into the white (normal) section. If the needle remains in the blue section, or enters the red (hot) section, a fault is indicated, and advice should be sought (refer to *'Fault finding'* on page 123).
   Do not continue to run an engine which shows signs of overheating.

VAUXHALL NOVA

CONTROLS AND EQUIPMENT 19

Clock

An analogue quartz clock is fitted to some models. To adjust the time, depress the knurled knob in the centre of the clock, and turn as required.

Oil pressure gauge

Fitted to some models, the oil pressure gauge can give early warning of trouble in the lubrication system. The scale ranges from 0 to 5 bars (0 to 71 lbf/in^2), and with the engine at normal temperature, the oil pressure must not drop below 2 bars (28.5 lbf/in^2) at high engine speeds.

Voltmeter

Supplementing the ignition warning light on some models, the voltmeter indicates the state of charge of the battery, and also shows if the charging system is functioning correctly. During starting, the battery voltage should not drop into the red zone, and during driving, the indicator should stay in the white zone. Should this not be the case, and assuming that the alternator drivebelt is not broken, have the charging system and battery checked by an expert.

Warning lights

These lights warn the driver of a fault, or inform the driver that a particular device is in operation.

BRAKE FLUID LOW LEVEL/HANDBRAKE 'ON' WARNING LIGHT

On some models, these two functions are on separate warning lights – the warning light with the letter 'P' lights up if the handbrake is applied with the ignition switched on, and the warning light with the exclamation mark lights up if the brake fluid drops below the minimum mark (with the ignition switched on). Where the two functions are combined, if the light comes on with the ignition switched on, check first that the handbrake is fully released, and if the light does not then go out, stop immediately and check the brake fluid level (refer to *'Regular checks'* on page 76).

Do not continue to drive the car if a brake fluid leak is suspected.

LOW OIL PRESSURE WARNING LIGHT

Warns that the engine oil pressure is low. If the light stays on for more than a few seconds after start-up, it's likely that the engine is worn, and advice should be sought. If the light comes on whilst driving, **switch off** the engine immediately and seek advice. When the engine is hot, the light may come on intermittently when idling, but it should go out when the engine speed is increased.

IGNITION WARNING LIGHT

Acts as a reminder that the ignition circuit is switched on, if the engine isn't running, and acts as a no-charge warning light if the engine is running. If the light stays on after starting, or comes on whilst driving, the battery is not charging properly. The battery may therefore become discharged, and the best course of action is to stop and seek advice. If the light comes on whilst driving on 1.0 litre overhead valve engine models, the alternator/water pump drivebelt may have broken. To continue driving would cause the engine to overheat, because the water pump would not be turning.

MANUAL CHOKE WARNING LIGHT

Acts as a reminder that the choke is in operation. The light will go out when the choke control is pushed in.

HEADLIGHT MAIN BEAM WARNING LIGHT

Acts as a reminder that the headlight main beam is switched on. Will also light up when the headlamp flasher is operated.

DIRECTION INDICATOR WARNING LIGHT

Shows that the direction indicators are switched on. Rapid flashing of the warning light means that a bulb has blown.

TRAILER INDICATOR WARNING LIGHT

With the trailer or caravan electrics hooked up, this warning light should flash in unison with the car indicator warning light. Failure to do so indicates that a bulb has blown on the trailer lighting, or that there is a bad connection at the trailer plug and socket.

CONTROLS AND EQUIPMENT

ENGINE CONTROL INDICATOR LIGHT

This indicates any fault in the electronic engine management system. It lights up when the engine is first switched on, and remains on during starting, but should go out when the engine is running. If the warning light comes on during driving, a fault has occurred and the electronic system automatically switches to its emergency running programme. The car may be driven for a short distance with the warning light on, but continued driving may result in increased fuel consumption and lack of response from the engine. A Vauxhall dealer should be consulted as soon as possible. If the warning light comes on briefly and then goes out again, this is of no significance.

Ignition switch/steering lock

The switch has four positions as follows:
- **B** Ignition off, steering locked
- **I** Ignition off, accessory circuits on, steering unlocked
- **II** Ignition on, and all electrical circuits on
- **III** Starter motor operates (release the key immediately the engine starts)

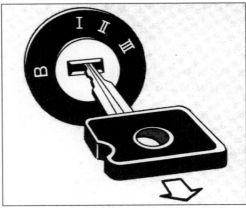

▲ *Ignition switch/steering lock positions. Withdraw the ignition key in position **B** to lock the steering*

To lock the steering, the ignition key must be withdrawn in position **B** – when the steering wheel is turned slightly, the lock will engage. To release the lock, insert the ignition key and turn it to position **1**, at the same time rocking the steering wheel slightly from side to side.

Gearbox

MANUAL GEARBOX

Either a 4 or 5-speed gearbox may be fitted. The gear positions follow the usual 'H' pattern, with 5th gear to the right of the 3rd gear position, and reverse gear to the left of the 1st gear position.

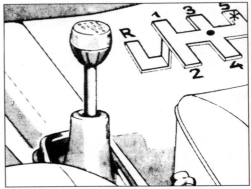

▲ *Gear lever positions*
● *Indicates Neutral position*

To engage reverse gear, move the gear lever to the left, then press it downwards to the left and forwards. Only engage reverse gear with the vehicle stationary, and wait at least 3 seconds after depressing the clutch pedal.

To engage 5th gear, move the gear lever to the right against the resistance, then move it forwards.

Left-hand multi-function switch

Controls the direction indicators, main and dipped beam, and headlight flash. Move the switch upwards to signal a right-hand turn, and downwards to signal a left-hand turn.

The switch will automatically cancel when the steering wheel is returned to the straight-ahead position. When lane-changing or making minor steering movements, the cancel function will not operate, and for indication in these circumstances, the switch should be moved part-way to the first stop and held in this position by hand. The indicators will flash until such time as the switch is released.

The main beam position is selected by

VAUXHALL NOVA

CONTROLS AND EQUIPMENT

pushing the switch away from the steering wheel, and the dipped beam position by pulling the switch towards the steering wheel to the next position.

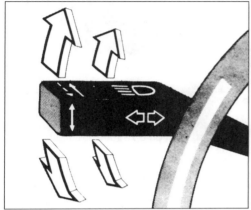

▲ Direction indicator selection

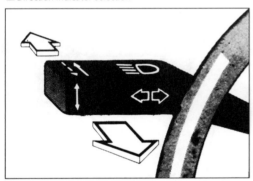

▲ Main and dipped beam selection

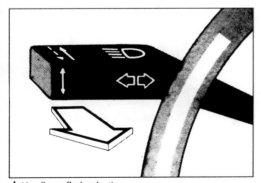

▲ Headlamp flash selection

To operate the headlamp flash, pull the switch towards the steering wheel against the resistance.

Releasing the switch will turn the headlamp flash off. The headlamp flash may also be operated together with the direction indicators.

Exterior and interior lights switch

Controls the sidelights/tail lights/number plate lights and headlights, and also the interior lights.

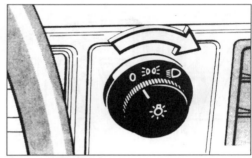

▲ Exterior lights switch

Turning the switch to the first position will operate the sidelights, and on models fitted with dim-dip lighting (August 1986 on), the headlights will glow dimly when the ignition is switched on. On later models, if the driver's front door is opened with the switch in either the sidelight or main beam positions, and the ignition key either removed or in the **B** or **I** positions, a warning buzzer will operate. Turning the switch to the second position will operate the headlights.

▲ Interior lights operation

VAUXHALL NOVA

CONTROLS AND EQUIPMENT

To operate the interior lights, the switch knob is pulled outwards.

The interior lights also come on automatically when the front doors are opened, and go out when the doors are closed.

Headlamp range adjustment control

On certain models produced from November 1990 onwards, a headlamp range adjustment control is fitted to the facia, just to the right of the steering wheel.

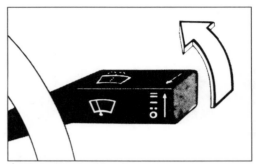
▲ Windscreen wiper operation

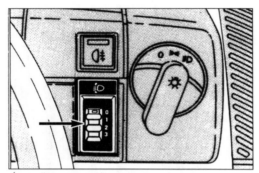
▲ Headlamp range adjustment control

The control has four settings as follows.
0   Driver's seat occupied
1   All seats occupied
2   All seats occupied and load in luggage compartment
3   Driver's seat occupied and load in luggage compartment

Right-hand multi-function switch

Controls the windscreen wipers and washers. Where a headlight wash/wipe system is fitted, this will operate whenever the screen washer is operated with the sidelights or headlights switched on. On models from November 1990 onwards, the switch also operates the tailgate wash/wipe system. On models produced before this date, the rear wash/wipe system is operated by a switch on the facia.

To operate the windscreen wipers, move the switch upwards from its 'off' position.

Where the switch has four positions, the first position will operate the wipers intermittently, pausing for several seconds between each wipe. The second position on a four-position switch (or the first position on a three-position switch) will operate the wipers continuously at a slow speed, and the top position will operate them continuously at a fast speed. To operate the washers, pull the switch towards the steering wheel; release the switch to stop the washers.

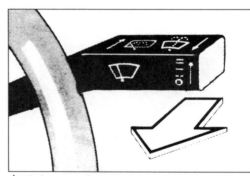
▲ Windscreen washer operation

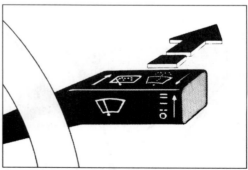
▲ Rear wipe/wash operation for models from November 1990 on

CONTROLS AND EQUIPMENT

The wipers operate automatically for several wipes while the washers are operating, and on models so equipped, the headlight wipers and washers will operate at the same time if the sidelights or headlights are switched on.

On November 1990-on models fitted with a rear wiper, push the switch away from the steering wheel to the first stop to operate the rear wiper intermittently, and to the second stop to cause the washers to operate automatically as the rear wiper is in operation.

Tailgate wash/wipe switch

This facia-mounted switch is fitted to certain models before November 1990. Raise the switch to its first position to operate the rear wiper, then raise it against the spring pressure to the second position to operate the washers.

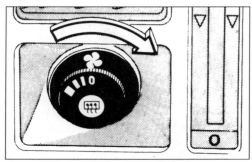
▲ Heater blower control knob (early facia shown)

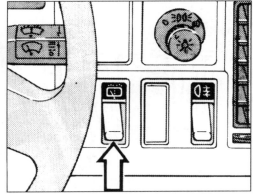

▲ Tailgate wipe/wash switch operation for models before November 1990

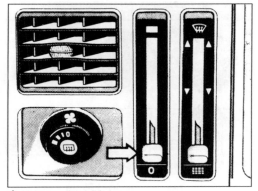
▲ Air temperature slide lever

Heating and ventilation controls

The rotary knob located beneath the right-hand centre air vent operates the heater blower.

Positioned fully anti-clockwise the blower is switched off, and turning it clockwise operates the blower at one of two or three speeds according to model, from slow to fast.

On models up to the end of October 1990, two sliders are provided for controlling the air temperature and distribution. The left-hand heater slide lever controls the air temperature. Maximum heat is obtained with the lever fully upwards.

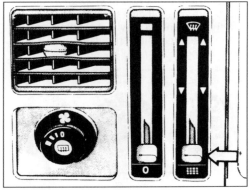
▲ Air distribution slide lever

Since the heater obtains its heat from the engine coolant, the engine must have been running for several minutes before any heat output can be expected. The right-hand heater slide lever controls the air distribution. Air is

CONTROLS AND EQUIPMENT

directed to the windscreen for defrosting in the upper position, to the foot area in the centre position, and to the head area in the bottom position. Between the top and middle positions, the air direction is variable between the windscreen and foot area.

On models from November 1990 onwards, the air temperature and distribution is controlled by rotary knobs each side of the heater blower knob. The left-hand knob controls the temperature – the coldest position is with the knob turned fully anti-clockwise, and maximum heat is with it turned fully clockwise.

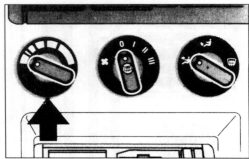

▲ *Air temperature control knob*

Air distribution is controlled by the right-hand knob – turned fully clockwise, air is directed onto the windscreen for defrosting. The centre position directs air to the foot area, and the fully anti-clockwise position directs air to the head area. Air distribution is variable between positions.

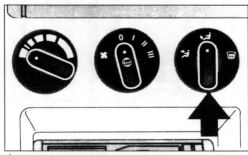

▲ *Air distribution control knob*

On all models, fresh unheated air can be directed into the vehicle through the centre vents.

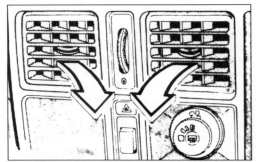

▲ *Central air vents*

The air supply is regulated by means of the control wheel (starting from position **0**), and the air direction can be controlled by tilting the vents and adjusting the fins as required. The air volume is controlled by the heater blower. The side vents operate in a similar manner.

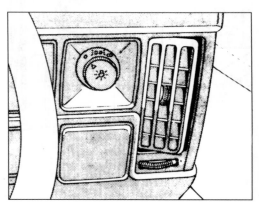

▲ *Side air vent*

▲ *Door window demister vent on early models*

VAUXHALL NOVA

CONTROLS AND EQUIPMENT

On models before November 1990, the direction of air to the front door windows for demisting is controlled by moving the right-hand distribution lever upwards. Hot or cold air may be supplied depending on the position of the left-hand lever.

Hazard flasher switch

Operates all direction indicators in unison, regardless of the position of the ignition switch. On models up to the end of October 1990, the hazard flasher switch is located on the right-hand side of the clock, and on later models it is located above the clock. The switch remains lit all the time that the ignition is switched on, so that it is easily located. Depress the switch once to switch on the hazard lights, and depress the switch again to switch them off.

▲ Hazard flasher switch on models up to the end of October 1990

▲ Hazard flasher switch on models from November 1990

Rear foglight switch

Operates the rear foglights. On models up to the end of October 1990, the toggle switch is located on the right-hand side of the facia. On very early models, the switch is located on the left-hand side of the front foglight switch, but as from October 1988 the two switch positions were swapped around.

▲ Rear foglight switch on models up to the end of October 1990

On models from November 1990 on, the pushbutton switch is located to the right of the steering wheel.

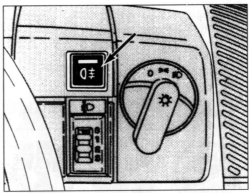

▲ Rear foglight switch on models from November 1990 on

On most models, the sidelights must be switched on for the rear foglights to operate, and on some models the rear foglights will go out when the headlamp main beam is switched on.

VAUXHALL NOVA

26 CONTROLS AND EQUIPMENT

Front foglight switch

Operates the front foglights, provided that the sidelights and ignition are switched on. On some models, the front foglights are extinguished when the headlamp main beam is switched on. Where fitted, the switch is located on the right-hand side of the facia.

▲ Front foglight switch on models up to the end of October 1990

Heated rear window switch

Operates the heated rear window, provided that the ignition is switched on.

The switch is incorporated in the heater blower rotary knob – pull the knob out to operate the heated rear window.

To avoid unnecessary electrical loads, it's recommended that the device is switched off as soon as demisting is complete.

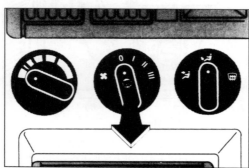

▲ Heated rear window switch operation (November 1990-on model shown)

Electric door mirror switch

On models fitted with electric door mirrors, the four-way controlling switch is located on the right-hand side of the facia next to the lighting switch.

▲ Four-way switch for controlling the electric door mirrors

Press on the edge of the switch to move the mirror in the desired direction.

Heated front seats switch

On certain models from November 1990 on, the front seat heating pushbutton switches are located on the centre console in front of the gear lever. The right-hand button operates the driver's seat, and the left-hand button operates the passenger's seat.

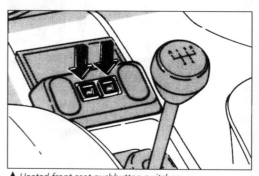

▲ Heated front seat pushbutton switches

Loudspeaker fader control

This feature is fitted to models up to the end of October 1990. The control determines the volume between the front and rear loudspeakers. With the knob in its central position, the volume is equal on all loudspeakers, but moving the knob to either side increases or decreases the front speaker volume in relation to the rear speakers.

VAUXHALL NOVA

CONTROLS AND EQUIPMENT

Bonnet release lever

Mounted beneath the right-hand side of the instrument panel; pull the lever to release the bonnet lock. At this stage, the bonnet will still be held by the safety catch.

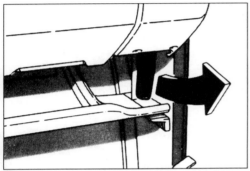

▲ Bonnet release lever under the right-hand side of the instrument panel

Lift the bonnet safety catch located below and slightly to the right of the centre of the bonnet (viewed from the front). Open the bonnet, and support it by releasing the stay on the front crossmember and locating it in the special slot under the right-hand corner of the bonnet.

▲ Safety catch under the centre of the bonnet

When closing the bonnet, make sure that the release lever inside the car is fully returned to its rest position. Check that the stay is firmly in its retainer, then lower the bonnet until it is approximately six inches above the lock and let it drop the final distance. Finally press on the bonnet to confirm that it is firmly locked.

▲ Support stay located in the special slot and in retainer

INTERIOR EQUIPMENT

Cigarette lighter

Operates regardless of the position of the ignition switch. Press the lighter in, then release it, and wait until it pops out ready for use.

Electric windows

On models fitted with electric windows, the operating switches are located on the centre console in front of the gear lever.

▲ Electric window switches on models up to the end of October 1990

The switches are of the rocker type – press the front edge of the switch to close the window, and the rear edge to open the window.

As a precaution against injury to children, remove the ignition key when leaving the car. On models from November 1990 on, the

CONTROLS AND EQUIPMENT

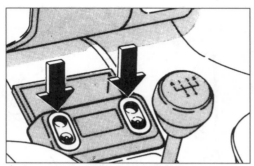

▲ Electric window switches on models from November 1990 on

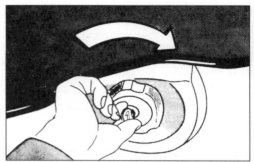

▲ Using a coin to turn the screw in the control knob

electric windows incorporate an overload switch which will cut off the supply to the electric motors for a short period in the event of an obstruction (such as stray fingers!) preventing the windows fully closing.

Sunroof

Three types of sunroof may be fitted. Up to August 1986, the sunroof was of the hinged removable type, and from this date onwards it was of the sliding type. On models from November 1990, the sliding roof may be electrically operated.

To raise or lower the hinged type sunroof, turn the control knob as required.

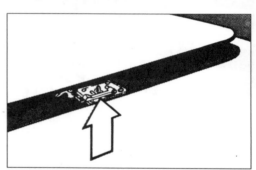

▲ Safety catch on the hinged sunroof

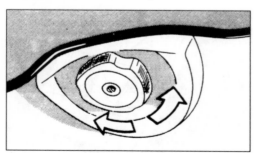

▲ Hinged sunroof operating control knob

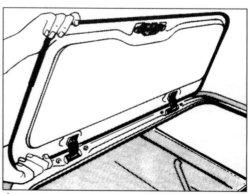

▲ Removing the hinged sunroof

To remove the sunroof, close it and turn the screw in the centre of the control knob clockwise using a coin or similar tool.

Raise the roof slightly and disengage the safety catch by pressing the release lever, then pivot it and remove it from its retainers.

A special storage pouch is provided for the roof panel, on the back of the rear seat.

Stow the panel with the front hinge tongues upwards, and the convex surface towards the back of the seat. When refitting the roof panel, turn the control knob to the open position. Fit the front hinge tongues into their retainers, engage the safety catch and close the roof using the control knob.

VAUXHALL NOVA

CONTROLS AND EQUIPMENT 29

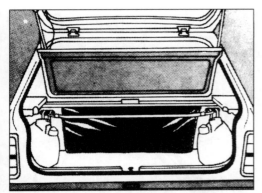

▲ Storage pouch for sunroof on back of rear seat (Hatchback model shown)

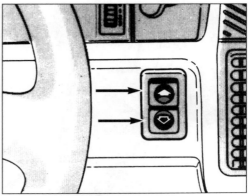

▲ Electric sunroof operating switches

The manually-adjusted sliding sunroof is operated by a cranked handle which, when turned, lifts the rear edge of the sunroof or moves it to the desired position. An adjustable sunshade is located beneath the sunroof, but this should not be pulled forward when the sunroof is open. If the sunshade is accidentally pushed right back, it can be pulled out again by fully opening the sunroof and closing it again.

To open the sliding roof, pull the cranked handle from its recess, then unlock it by pressing the button.

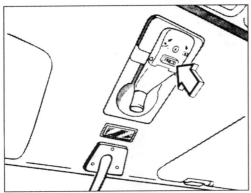

▲ Unlocking button on the cranked handle on sliding sunroof

Turn the handle anti-clockwise to move the sunroof rearwards to the desired position, then press the handle back into its recess to lock it. If it is only required to raise the rear edge of the sunroof for ventilation (ie without moving it rearwards), turn the handle clockwise, then press it back into its recess.

The electric sunroof is operated by two switches located beneath the lighting switch.

Pressing the bottom switch opens the roof, and pressing the top switch closes the roof. Release the switch as soon as the roof is fully closed. Do not open the roof when travelling at speeds in excess of 87 mph (140 kph), or when it is frozen or covered with snow. Make sure that there is no obstruction in the open roof space before closing the roof, and do not leave the ignition key in the car when children are left in it.

If a fault occurs in the electric sunroof, it may be operated using the emergency crank. Open the cover at the rear of the roof using the ignition key, then insert the crank and turn as necessary.

Dipping rear view mirror

Pull back the lever under the mirror to reduce the glare from the lights of following vehicles at night. Note that this feature is not fitted to all early models.

Seat belts

The operation of the seat belts is self-explanatory, but certain points require additional explanation.

FRONT SEAT BELT HEIGHT ADJUSTMENT

On certain models, the height of the upper seat belt mounting can be adjusted. To achieve

VAUXHALL NOVA

CONTROLS AND EQUIPMENT

▲ Front seat belt height adjustment

▲ Shortening the rear seat belt

this, depress the button and move the mounting to the required setting, allowing it to lock audibly into position.

REAR STATIC SEAT BELT ADJUSTMENT

Release the seat belt from the adjuster, and position it over the shoulder, checking that it is not twisted.

If the seat belt requires shortening, simply pull on the free end until it feels comfortable.

To lengthen the seat belt, push up the free end to provide extra belt, then tilt the adjuster on the side pillar and pull the main belt downwards until comfortable. Pull the free end down to eliminate the loose belt.

▲ Lengthening the rear seat belt

▲ Positioning the rear seat belt over the shoulder

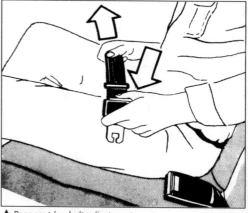

▲ Rear seat lap belt adjustment

VAUXHALL NOVA

CONTROLS AND EQUIPMENT

REAR SEAT BELTS

Where fitted, the centre rear seat belt is of the lap type (non-inertia), and must be manually adjusted by the wearer. To adjust its length, depress the black button above the lock tongue, and pull the webbing through as required.

Front seat adjustment

SEAT SLIDING CONTROL

To move the front seat backwards or forwards, pull up the lever located on its front, then slide the seat to the required position and release the lever to lock it.

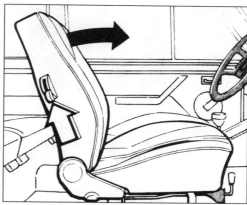
▲ Front seat backrest release

SEAT RECLINING CONTROL

Turn the wheel on the inside rear base of the seat to alter the angle of the seat back. Do not lean back on the seat while making the adjustment.

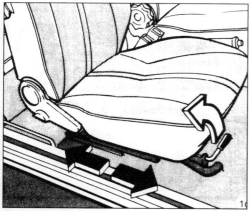
▲ Front seat position adjustment

On models up to the end of July 1986, it is possible to extend the rear stop of the front seat adjustment for tall people. Adjust the seat fully forwards and tilt the backrest forwards, then reposition the safety catch in the rearmost hole in the outer seat runner.

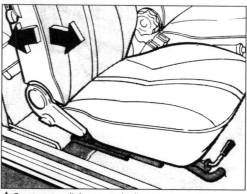
▲ Front seat reclining control adjustment

Adjustable head restraints

SEAT BACK RELEASE (TWO-DOOR MODELS)

Pull up the lever on the outside of the front seat backrest, and swivel the backrest forwards to allow rear seat passengers to enter.

The backrest will click into place when it is swivelled back to its normal position.

The head restraints can be adjusted for height – the upper edge should be approximately at eye height, not neck height. To make an adjustment, simply move the restraint to the required position – it will be locked in the vertical position automatically.

To remove the head restraints on models up to approximately 1984, use a screwdriver to prise out the retaining springs, then lift the restraint from the back of the seat.

VAUXHALL NOVA

CONTROLS AND EQUIPMENT

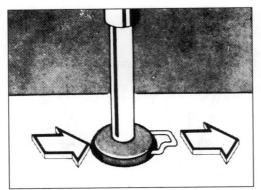

▲ Pull out the retaining spring to remove the head restraint

When refitting it, make sure that the spring is inserted with its bent side pointing towards the rear.

To remove the head restraints on models after approximately 1984, release the catch retaining springs by pushing them to the rear, then pull the restraint upwards.

Folding rear seats

To fold down the rear seat on Hatchback models, first secure the side seat belts in the special clips, then pull up or press down (according to model) the buttons on the top of the backrest, and fold the backrest forwards onto the cushion.

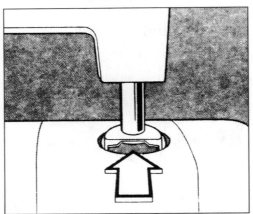

▲ Head restraint retaining spring on models from approximately 1984 to October 1990

▲ Rear seat positions for increasing luggage space

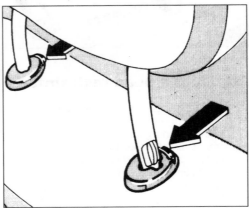

▲ Head restraint retaining spring on models from November 1990

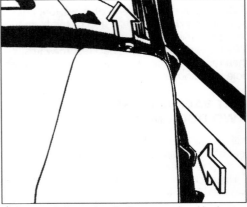

▲ Rear seat belt securing clip (lower arrow) and backrest release button (upper arrow)

VAUXHALL NOVA

CONTROLS AND EQUIPMENT

If necessary, the rear parcel shelf may be removed to provide additional space. To do this, push the locks on the rear side trim panels outwards to release the front of the shelf, then unhook the straps from the tailgate. The shelf may be stored behind the front seats.

▲ Rear parcel shelf locks

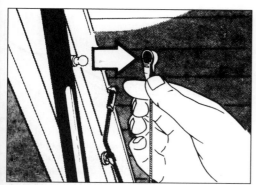
▲ Disconnecting the rear parcel shelf straps from the tailgate

Access to the rear luggage compartment may also be gained with the rear seat backrest in its normal position by releasing the two locks and lifting the front of the parcel shelf. Again, if tall objects are to be carried behind the rear seat, the rear shelf may be removed and stored behind the backrest.

To provide the maximum amount of space in the luggage compartment, first locate the rear seat belt buckles on the buttons provided on the backrest, then pull up the cushion using the centre loop, and fold the cushion forwards.

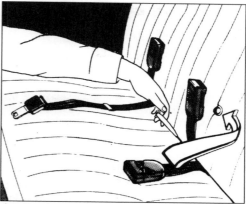
▲ Pulling the rear seat cushion forwards

Remove the rear parcel shelf and fold the rear seat backrest forwards as previously described.

When putting the rear seat backrest back in its normal position, make sure that it locks audibly in the catches. Before folding the cushion down, lift the lap seat belt ends, making sure that they are not twisted.

To remove the rear seat cushion on Saloon models, pull out the handles located under the front corners, then lift the cushion upwards and out of the car.

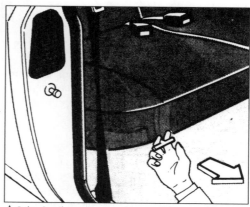
▲ Releasing the rear seat cushion on Saloon models

When refitting the cushion, push it fully under the backrest, then press down on each front corner until it locks into position.

VAUXHALL NOVA

EXTERIOR EQUIPMENT
Door locks

All doors may be locked from the inside by depressing the locking knob. On the passenger doors, the knob may be depressed with the door open so the door will be locked when it is closed. However, it is not possible to depress the knob on the driver's door with the door open; this is to prevent the driver from being locked out. To lock the driver's door, close the door, then turn the key in the lock.

Except on very early models, the rear door locks incorporate childproof catches, operated by pushing up the small lever on the rear edge of the door below the lock.

▲ *Unlocking the tailgate (early models) and bootlid*

▲ *Childproof catch on the rear doors*

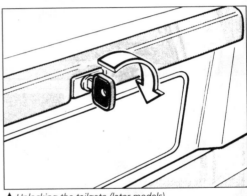

▲ *Unlocking the tailgate (later models)*

The rear doors cannot then be opened from inside the car, although they may be opened from outside the car, provided that the locking knob has not been depressed.

Tailgate/bootlid lock

To unlock either the tailgate or the bootlid, insert the key in the lock and turn clockwise. On early Hatchback models and all Saloon models, the tailgate/bootlid will be opened automatically; the key will return to the vertical position and may be removed.

On later Hatchback models, depress the button to open the tailgate. The key may be removed from its horizontal position, or turned to the vertical position and then removed.

After closing the tailgate (early models) or bootlid, the lock is automatically engaged, as the key cannot be removed from the unlocked position. On later Hatchback models where it is necessary to depress a button to open the tailgate, the key may be removed from either the locked or unlocked position. To decide whether the lock is locked or unlocked, look at the key slot – if it is vertical it is locked, and if it is horizontal it is unlocked.

Central door locking system

On models fitted with central locking, all doors are simultaneously locked when the driver's door is locked by turning the key in the lock, or by depressing the locking knob inside the car. This does not include the tailgate on Hatchback models or the bootlid on Saloon models, which must be locked separately.

ACCIDENTS AND EMERGENCIES

In the event of an accident, the first priority is safety. This may seem obvious, but in the heat of the moment, it's very easy to overlook certain points which may worsen the situation, or even cause another accident. The course of action to be taken will vary depending on how serious the accident is, and whether anyone is injured, but always try to think clearly, and don't panic.

We don't suggest that you consult this Section at the scene of an accident, but hopefully the following advice will help you to be better prepared to deal with the situation should you be unfortunate enough to appear on the scene of an accident or become involved in one yourself.

HOW TO COPE WITH AN ACCIDENT

Deal with any possible further danger

Further collisions and fire are the dangers in a road accident.

- **If possible warn other traffic**

    Where possible, switch on the car's hazard warning flashers.

    If a warning triangle is carried, position it a reasonable distance away from the scene of the accident, to give approaching drivers sufficient warning to enable them to slow down. Decide from which direction the approaching traffic will have least warning, and position the triangle accordingly.

    If possible, send someone to warn approaching traffic of the danger which exists, and to encourage the traffic to slow down.

- **Switch off the ignition and impose a 'No Smoking' ban**

    This will reduce the possibility of fire, should there be a petrol leak.

Call for assistance

Send someone to call the emergency services (dial 999), and make sure that all the necessary information is provided to the operator. Give the exact location of the accident, and the number of vehicles and if applicable the number of casualties involved. Refer to the *'Motorway breakdowns'* Section on page 44 for details of how to call for assistance on a motorway.

- **Call an ambulance**

    If anyone is seriously injured or trapped in a vehicle.

- **Call the fire brigade**

    If anyone is trapped in a vehicle, or if you think that there is a risk of fire.

- **Call the police**

    If any of the above conditions apply, or if the accident is a hazard to other traffic. In most cases, the accident must be reported to the police within 24 hours even if none of the above conditions apply (refer to *'Requirements of the law in the event of an accident'* on page 38).

Administer first aid

Refer to *'First aid'* on page 36.

Provide your details

If you are involved in the accident as a driver or car owner, provide your personal and vehicle details to anyone having reasonable grounds to ask for them. Also inform the police of the details of the accident as soon as possible, if not already done.

VAUXHALL NOVA

FIRST AID

Always carry a first aid kit

If possible, learn first aid by attending a suitable course – contact the St John Ambulance Association or Brigade, St Andrew's Ambulance Association or the British Red Cross Society in your local area for details. These organisations will also be able to provide further training in heart massage and mouth-to-mouth resuscitation which could enable you to save someone's life in an emergency.

Before proceeding with any kind of first aid treatment, deal with any possible further danger, and call for assistance, as described previously in this Chapter.

To help remember the sequence of action which should be followed when dealing with a seriously injured casualty, use the **ABC** of emergency first aid.

Casualties remaining in vehicles

Any injured people remaining in vehicles should not be moved unless there is a risk of further danger (such as from fire, further collisions etc).

Casualties outside the vehicles

Recovery position

To prevent the possibility of an unconscious but breathing casualty choking or suffocating, he/she must be placed in the recovery position.

▲ *The recovery position*

Lie the casualty on his/her side. Bend the leg and position the arms as shown to support the head.

First aid for other injuries

Always treat injuries in the following order of priority:
- Ensure that the casualty has a clear airway. If not treat as described opposite.
- Put any breathing, unconscious casualty into the recovery position as described previously.
- Treat any severe bleeding by applying direct pressure to the wound. Maintain the pressure for at least 10 minutes and apply a suitable clean dressing. If the wound continues to bleed, apply further pressure and dressings **over** that already in place. If bleeding from a limb, and as long as the limb is not broken, lift the affected limb to reduce the bleeding.
- Broken limbs should not be moved, unless the casualty has to be moved in order to avoid further injury. Support the affected limb(s) by placing blankets, bags, etc, alongside.
- Burns should be treated with plenty of cold water as soon as possible. Don't attempt to remove any clothing, but cover with a clean dressing.

Reassurance

The casualty may be in shock, but prompt treatment will minimise this. Reassure the casualty confidently, avoid unnecessary movement, and keep him/her comfortable and warm. Make sure that the casualty is not left alone.

Give the casualty NOTHING to eat, drink or smoke.

ACCIDENTS AND EMERGENCIES 37

Airway
- **Check for breathing**

Check that the airway is clear. If the casualty is breathing noisily or appears not to be breathing at all, remove any obvious obstruction in the mouth and tilt the head back as far as possible. Breathing may then start.

Breathing
- **If breathing has stopped**

If after clearing the airway the casualty still appears not to be breathing, look to see if the chest or abdomen is moving. Place your ear close to the casualty's mouth to listen and feel for breathing, and look to see if there's any movement of the chest. If you can't detect anything, you must start breathing for the casualty (mouth-to-mouth resuscitation).

To do this, pinch the casualty's nostrils firmly, keep his/her chin raised, and seal your lips around the casualty's mouth. Breathe out through your mouth into the casualty until the chest rises, then remove your mouth and allow the casualty's chest to fall. Give one further breath.

If giving mouth-to-mouth resuscitation to a child, remember that an adult's lungs are significantly larger than those of a child. Care must be taken not to overinflate a child's lungs, as this may cause injury.

- **Check for a pulse at the base of the neck**

If the heart is beating, continue to breathe into the casualty at a rate of one breath every 5 seconds, until he/she is able to breathe unaided, then place the casualty in the recovery position (refer to *'Recovery position'* opposite).

Circulation
- **If the heart has stopped**

If there is no pulse at the neck, then external chest compression (heart massage) must be started.

To do this, if the casualty is still in the vehicle, he/she must be removed and laid on his/her back on the ground. Kneel down on one side of the casualty. Feel for the lower half of the casualty's breastbone and place the heel of one of your hands on this part of the bone, keeping your fingers off the casualty's chest. Cover this hand with your other hand, interlocking your fingers. Keeping your arms straight and vertical, press the breastbone down about 4 to 5 centimetres (1½ to 2 inches). The pressure should be smooth but not jerky. Do this 15 times, at the rate of just over once a second. It may help you to count aloud as you press.

Follow this by 2 breaths as described under *'Breathing'*.

Repeat the cycle of 2 breaths followed by 15 compressions, twice more, ending with 2 breaths, then re-check the pulse.

- **If there is still no pulse**

Repeat the cycle 9 times and then check the pulse. Continue this procedure until there is a pulse or until professional help arrives.

VAUXHALL NOVA

ACCIDENTS AND EMERGENCIES

REQUIREMENTS OF THE LAW IN THE EVENT OF AN ACCIDENT

The following is taken from the Road Traffic Act of 1988.

If you are involved in an accident – which causes damage or injury to any other person, or another vehicle, or any animal (horse, cattle, ass, mule, sheep, pig, goat or dog) not in your vehicle, or roadside property:

You must
- stop;
- give your own and the vehicle owner's name and address and the registration mark of the vehicle to anyone having reasonable grounds for requiring them;
- if you do not give your name and address to any such person at the time, report the accident to the police as soon as reasonably practicable, and in any case within 24 hours;
- if anyone is injured and you do not produce your certificate of insurance at the time to the police or to anyone who has with reasonable grounds required its production, report the accident to the police as soon as possible, and in any case within 24 hours, and either produce your certificate of insurance to the police when reporting the accident or ensure that it is produced within seven days thereafter at any police station you select.

ESSENTIAL DETAILS TO RECORD AFTER AN ACCIDENT

If you're involved in an accident, note down the following details which will help you to complete the accident report form for your insurance company, and will help you if the police become involved.
- The name and address of the other driver and those of the vehicle owner, if different.
- The name(s) and address(es) of any witness(es) (independent witnesses are particularly important).
- A description of any injury to yourself or others.
- Details of any damage caused to the vehicles involved or other property.
- The name and address of the other driver's insurance company and, if possible, the number of his/her certificate of insurance.
- The registration number of the other vehicle (check this against the tax disc if possible).
- The number of any police officer attending the scene.
- The location, time and date of the accident.
- The speed of the vehicles involved.
- The width of the road, details of road markings and signs, the state of the road surface, and the weather conditions.
- Any marks or debris on the road relevant to the accident.
- A rough sketch showing the vehicle positions before and after the accident. It's helpful to make a note of the vehicle positions in terms of distance from fixed landmarks, such as lamp posts, buildings, etc.
- Whether any of the other vehicle occupants were wearing seat belts.
- If the accident occurred at night or in poor visibility, whether vehicle lights or street lights were switched on.
- If you have a camera, take a picture of the vehicles and the scene.

If the other driver refuses to give you his/her name and address, or if you consider that he/she has committed a criminal offence, inform the police immediately.

VAUXHALL NOVA

ACCIDENT REPORT DETAILS   39

Use this page to record all the relevant details in the event of an accident.
You should transfer all the information recorded on this page onto the Motor Vehicle Accident Report form which you can obtain from your insurance company.

OTHER DRIVER DETAILS

FULL NAME OF DRIVER:

ADDRESS:

POST CODE:

HOME TELEPHONE: WORK TELEPHONE:

DRIVING LICENCE NUMBER: ISSUED:

DATE OF BIRTH: DATE DRIVING TEST PASSED:

TYPE OF LICENCE: (tick box) Full ☐ Provisional ☐ Heavy Goods ☐

PERMITTED GROUPS:

FULL NAME OF OWNER:

ADDRESS:

POST CODE:

TELEPHONE:

INSURANCE COMPANY:

POLICY NUMBER:

AGENT OR BROKER:

TELEPHONE:

OTHER VEHICLE DETAILS

MAKE: MODEL:

YEAR: CC:

REGISTRATION NUMBER:

V A U X H A L L   N O V A

40 ACCIDENT REPORT DETAILS

CIRCUMSTANCES OF ACCIDENT

DATE: TIME: am/pm

PLACE: (Street or Road)

TOWN: COUNTY:

SPEED: WERE THE POLICE CALLED? Yes ☐ No ☐

If 'Yes' give details of Police Constabulary concerned:

DETAILS OF WHAT HAPPENED (Please use the page opposite to make a rough sketch map)

WHAT WAS THE WEATHER LIKE?

INDEPENDENT WITNESS 1

TELEPHONE:

INDEPENDENT WITNESS 2

TELEPHONE:

DETAILS OF ANY INJURIES SUSTAINED BY EITHER PARTY:

VAUXHALL NOVA

ACCIDENT REPORT SKETCH MAP   41

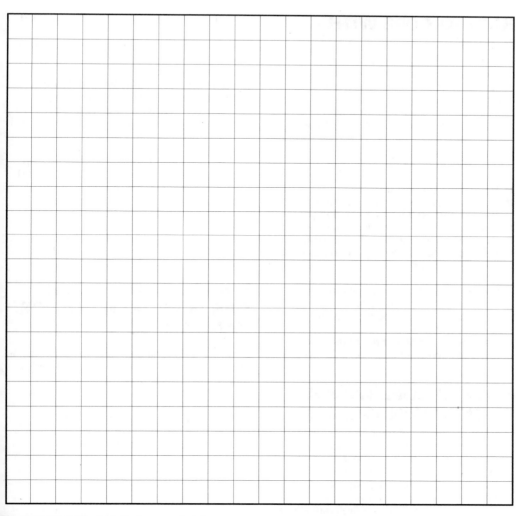

Essential items to include in your sketch map:
- The layout of the road and its approaches.
- The directions and identities of both vehicles.
- Their relative positions at the time of impact.
- The road signs and road markings.
- Names of the Streets and Roads.

VAUXHALL NOVA

ACCIDENTS AND EMERGENCIES

HOW TO COPE WITH A FIRE

Fire is unpredictable, and particularly dangerous when cars are involved because of the presence of petrol, which is highly inflammable.

It's sensible to carry a fire extinguisher on board your car, but bear in mind that the average car fire extinguisher is suitable for use only on the smallest of fires.

In the event of your car catching fire:

- **Switch off the ignition.**
- **Get any passengers and yourself out of the car and well away from danger.**
- **Impose an immediate 'No Smoking' ban.**
- **DO NOT expose yourself, or anyone else, to unnecessary risk in an attempt to control the fire.** Minor fires can be controlled using a suitable extinguisher, but always use an extinguisher at arms-length, and bear in mind that a small vehicle fire can develop into a serious situation without warning.
- **Call for assistance, if necessary** – refer to 'How to cope with an accident' on page 35.

HOW TO COPE WITH A BROKEN WINDSCREEN

There are a number of national specialist companies who offer roadside assistance to drivers who suffer broken windscreens, and it's a good idea to carry the 'phone number of a suitable specialist in your car for use in such situations. Some car insurance policies enable the use of such companies at preferential rates, and it's worth enquiring about this when obtaining insurance quotes.

- **Most of the windscreens fitted to modern cars are of the laminated type.** Laminated windscreens are made from layers of glass and plastic (usually two layers of glass sandwiching a layer of plastic) which prevents the windscreen from shattering, and preserves the driver's vision in the event of the screen suffering an impact.
  A sharp impact may crack the outer glass layer, but clear vision is usually maintained.

Obviously, if the windscreen is damaged, it should be replaced at the earliest opportunity, but there is no need to curtail your journey unless the damage is particularly serious.

- **Some older cars may be fitted with toughened windscreens.** If a toughened windscreen suffers an impact, the glass will normally stay intact, but severe 'crazing' will usually occur which is likely to seriously affect the driver's vision.
  If a toughened windscreen breaks, stop immediately, and seek assistance. **DO NOT** attempt to knock the broken glass out of the windscreen frame, as this is likely to result in injury to yourself, and damage to the car. Driving without a windscreen is extremely dangerous and should not be attempted.

WHAT TO DO IF YOUR CAR IS BROKEN INTO OR VANDALISED

If you are unfortunate enough to have your car broken into or vandalised, do not attempt to drive your car away from the scene unless you are satisfied that no damage has been caused which could affect the car's safety.

Where possible notify the police before moving your car, and inform your insurance company at the earliest opportunity.

VAUXHALL NOVA

BREAKDOWNS

43

In the unfortunate event of a breakdown, the first priority must always be safety – never risk causing an accident by attempting to move the car, or by trying to work on it if it's in a dangerous position.

Joining one of the national motoring organisations such as the AA or the RAC can provide you with a recovery and assistance service should you break down. It may also save you money, as local garages often charge high rates to provide a recovery service.

This Section provides advice on the course of action to follow should you break down. There are also Sections covering the procedure to follow for towing, and on how to cope with two of the more common causes of breakdowns – a puncture, and a flat battery.

Breakdowns on an ordinary road (not a motorway)

Most importantly of all, common sense must be used, but the following advice should prove helpful.

- **Warn approaching traffic,** to minimise the risk of a collision.
  Where possible, switch on the car's hazard warning flashers, and at night leave the sidelights switched on.
  If a warning triangle is carried, position it a reasonable distance away from the scene of the breakdown, to give approaching drivers sufficient warning to enable them to slow down. Decide from which direction the approaching traffic will have least warning, and position the triangle accordingly. Place the warning triangle on the road surface, out from the edge of the road, where it can be easily seen.
  If possible, send someone to warn approaching traffic of the danger which exists, and to encourage the traffic to slow down.
- **Get the passengers out of the car** as a precaution should the car be hit by another vehicle. The passengers should move well away from the car and approaching traffic, up (or down) an embankment, or into a nearby field for example.
- **If possible, move the car to a safe place.** If the car cannot be safely driven or pushed to a place of safety, call a recovery service to provide assistance. *Don't expose yourself or anyone else to unnecessary danger in order to move the car.*
- **If the car can't be safely moved, turn the steering wheel towards the side of the road.** If the car is hit, it will then be pushed into the side of the road, not into the path of traffic.
- **Try to find the cause of the problem.** Only attempt this if the car is in a safe position away from traffic. Refer to *'Fault finding'* on page 123 for details.
- **If necessary, call for assistance** from one of the national motoring organisations if you're a member, or from a local garage who can provide a recovery service.

VAUXHALL NOVA

BREAKDOWNS

Motorway breakdowns

In the event of a motorway breakdown, due to the speed and amount of traffic there are a few special points to consider, and the following advice should be followed.

- **If possible, switch on the car's hazard warning flashers, and at night leave the sidelights switched on.**
- **DO NOT open the doors nearest to the carriageway, and DO NOT stand at the rear of the car, or between it and the passing traffic.**
- **Get the car off the carriageway and onto the hard shoulder as quickly as possible, and as far to the left as possible** – never forget the danger from passing traffic. Turn the steering wheel to the left, so that the car will be pushed away from the carriageway if hit from behind.
- **If you can't move the car off the carriageway** – get everyone out of the car and well away from the carriageways (up/down an embankment, or into a nearby field for example) as quickly as possible. Never forget the danger from passing traffic.

  Call for assistance as described below.
  Attempt to warn approaching traffic of your car's presence by signalling from well to the back of the hard shoulder, NOT the carriageway. DO NOT put yourself at risk. Face the approaching traffic, and be prepared to move quickly clear of the hard shoulder onto the verge if any vehicles pull onto the hard shoulder.
- **Get the passengers out of the car** – as a precaution should the car be hit by another vehicle. The passengers should move well away from the car and approaching traffic, up (or down) an embankment, or into a nearby field for example. If animals are being carried, it may be advisable to leave them in the vehicle. In any case, animals and children **must** be kept under tight control.
- **Call for assistance** – using the nearest emergency telephone on your side of the carriageway. NEVER cross the carriageways to use the emergency telephones.

  The direction of the nearest telephone is indicated by an arrow on the marker post behind the hard shoulder (the marker posts are positioned 100 metres/300 feet apart).

Each telephone is coded with details of its location, and the operator will automatically be informed of your position on the motorway when you make the call.

Give the operator details of your car type and registration number and, if you're a member of one of the motoring organisations, ask the operator to arrange assistance.

- **Don't leave your car unattended for a long period.**

Changing a wheel

- **First of all, make sure that the car is in a safe position.** It may be safer to drive the car slowly forwards and risk damaging the wheel, than to risk causing an accident.
- **If you're in any doubt** as to whether you can safely change the wheel without putting yourself or other people at risk, call for assistance.
- **All the tools required to change a wheel, together with the spare wheel itself, are located beneath a cover in the rear luggage compartment.** The spare wheel is secured by a plastic wing nut, and the jack and tools are located under the spare wheel. If you're carrying any luggage, you'll need to move it first for access to the spare wheel. Lift the cover, unscrew the plastic wing nut, then lift the spare wheel, jack and tools out of the storage well. Place them on the ground (don't put any luggage back yet, because the wheel with the punctured tyre will have to be put back in the spare wheel well).

▲ *Spare wheel and tools location*

BREAKDOWNS 45

- **Make sure that the handbrake is applied,** and select reverse gear.
- **Prise off the wheel bolt caps and/or hub cap** with the screwdriver supplied in the tool kit – on models where the hub cap covers the complete main part of the wheel, use your fingers also to apply extra force to the edge of the cap. Where holes or slots are provided, insert your fingers in them and pull.

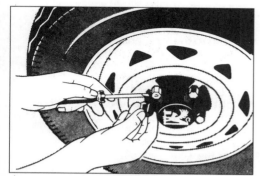
▲ *Removing the wheel bolt plastic caps*

▲ *Removing a hub cap with holes*

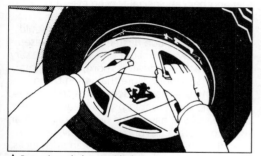
▲ *Removing a hub cap with slots*

- **Slacken the four wheel bolts** by at least half a turn using the wheelbrace provided. If possible, chock the diagonally-opposite wheel to the one being removed with large stones or pieces of wood.

▲ *Loosening the wheel bolts (plastic bolt caps already removed)*

- **The jacking points** are shown in the accompanying illustration and they are indicated on the car by two recesses on the side sills (below the doors), one in front of the rear wheel, and the other behind the front wheel.

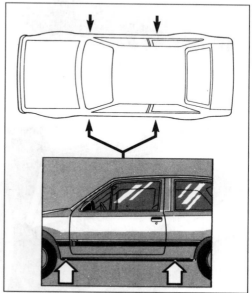
▲ *Jacking point locations (note that as from September 1988, the jacking points are slightly further to the rear than those indicated in this illustration)*

VAUXHALL NOVA

BREAKDOWNS

- **Choose the jacking point nearest the wheel to be changed,** and position the pad on the jack leg directly below it. Turn the jack handle clockwise so that its head moves upwards, and guide it onto the jacking point so that the vertical lip on the sill locates in the groove in the jack claw.

  Make doubly sure that the base of the jack is positioned directly below the jacking point, then continue to turn the handle clockwise. If the jack tilts at all or slips, lower it and start again. Carry on turning the jack handle slowly to raise the car until the wheel is clear of the ground, making sure that the car doesn't move on the jack. Slide the spare wheel under the sill for the time being, as near to the punctured wheel as possible. This will break the car's fall (and minimise the risk of injury to you) if the car slips off the jack, or the jack collapses.

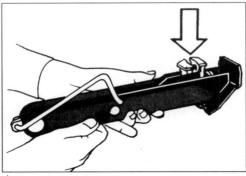

▲ Groove in the jack claw

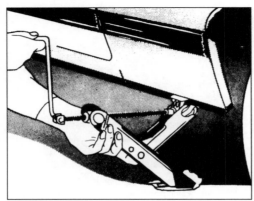

▲ Using the car jack

- **Unscrew all the wheel bolts,** then remove the wheel. Sometimes it may be necessary to tap the wheel in order to release it, but if this is done, make sure that the car doesn't move on the jack. Remove the spare wheel from under the side of the car, and slide the punctured wheel under there in its place for the time being.
- **Fit the spare wheel to the car,** and lightly tighten the wheel bolts in a diagonal sequence. Slide out the punctured wheel and slowly lower the car to the ground.
- **Remove the jack, and fully tighten the wheel bolts.** There's no need to strain yourself doing this, just make sure that the bolts are tight, using only the wheelbrace provided.
- **Refit the hub cap and/or wheel bolt caps,** giving a sharp tap with the palm of the hand to secure. Where applicable, make sure that the special hole for the valve is located over the valve – on some models a valve symbol is shown on the rear of the hub cap. On light alloy wheels, insert the pin on the rear of the hub cap in the hole in the rim.
- **Stow the punctured wheel and the tools** in the spare wheel well, and tighten the plastic wing nut.
- **Remove the wheel chocks,** and make a final check to ensure that all tools and debris have been cleared from the roadside.
- **Select neutral** before starting up and driving away.
- **Have the puncture repaired** at the earliest opportunity – another puncture will leave you stranded if you don't!

Towing

If your car needs to be towed, or you need to tow another car, special towing eyes are provided at the front and rear of the car.

When being towed, the ignition key should be turned to position **II** (steering lock released, and ignition warning light on). This is necessary for the direction indicators, horn, brake lights and wipers to work. Note that when being towed, the brake servo does not work (because the engine is not running). This means that the brake pedal will have to be pressed harder than usual to operate the brakes, so allowance should be made for greater stopping distances. If the breakdown doesn't prevent the engine

BREAKDOWNS 47

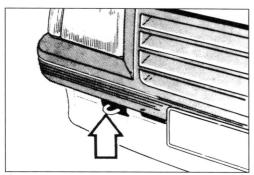

▲ *Front towing eye*

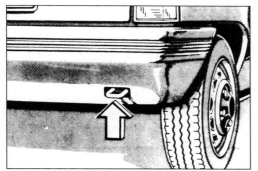

▲ *Rear towing eye*

from running, the engine could be started and allowed to idle so that the servo operates while towing. The gear lever should be positioned in neutral.

An 'On tow' notice should be displayed prominently at the rear of the car on tow to warn other drivers.

Make sure that both drivers know details of the route to be taken before moving off, as it is difficult, and dangerous, to try to communicate once under way.

Before moving away, the tow car should be driven slowly forwards to take up any slack in the tow rope.

The driver of the car on tow should make every effort to keep the tow rope tight at all times, by gently using the brakes if necessary, particularly when driving downhill.

Drive smoothly at all times, particularly when moving away from a standstill. Allow plenty of time to slow down and stop when approaching junctions and traffic queues.

VAUXHALL NOVA

Starting a car with a flat battery

Apart from old age (most batteries should last for at least three years), a battery will normally only go flat if there is a fault in the charging circuit (indicated by a continuously-glowing ignition warning light), or when a particular circuit (eg headlights) is left on for a long time with the engine switched off.

To start a car with a flat battery, you can either use a push or tow start (but not on models fitted with a catalytic converter), or a set of jump leads. Note, however, that where a maintenance-free battery is fitted, certain precautions must be observed. Batteries of this type were fitted extensively to the Nova range as original equipment, and will typically be marked 'Freedom Battery'. If the battery is of this type, look carefully on the top of the battery for a small round window – this indicates battery condition by changing colour, as shown in the accompanying illustration.

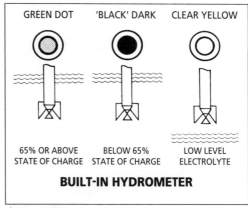

▲ *Battery condition indicator on maintenance-free battery*

Do not attempt to start the engine if the battery condition indicator is clear or yellow in colour. This means that the electrolyte level in the battery is too low to allow further use, and the battery should be renewed. If the indicator is either green (battery in good state of charge) or black (battery needs charging), start the engine using any of the following methods.

TOW/PUSH STARTING

Warning: *Due to the possibility of damaging the catalyst unit, models fitted with a catalytic converter should not be tow- or push-started – use the jump lead method described later in this Section.*

The method used to start the engine is the same whether the car is being towed or being pushed but, if the car is to be towed, first read the previous Section on towing for details of where to connect the tow rope.

Proceed as follows:

1 Turn the ignition key to position **II** to switch on the ignition and unlock the steering. Switch off all other electrical loads.

2 On models with a manual choke, if the engine is cold, pull out the choke control as far as it will go, and do not depress the accelerator pedal. If the engine is warm, do not pull out the choke control – hold the accelerator pedal depressed halfway to the floor. If the engine is hot, do not pull out the choke control – hold the accelerator pedal depressed fully.

3 On models with an automatic choke, fully depress the accelerator pedal once, then release it. If the engine is warm on these models, hold the accelerator pedal halfway down – do not fully depress it.

4 On fuel injection models, do not depress the accelerator pedal at all.

5 Depress the clutch pedal and select second or third gear. Hold the clutch pedal down.

6 Tow or push the car, then release the clutch pedal. The engine will turn over, and should start (don't worry about the car 'juddering' as the engine turns). If the car is being towed, as soon as the engine starts, depress the clutch pedal and *gently* brake the car so that you don't run into the tow vehicle.

7 When the engine is running, on models with an automatic choke, briefly depress the accelerator pedal – this will release the automatic choke mechanism and allow the engine to run at a slightly lower speed. On models with a manual choke, push the choke control in a little so that the engine still runs smoothly.

STARTING USING JUMP LEADS

Note: *Starting using jump leads can be dangerous if done incorrectly. If you're uncertain about the following procedure, it's recommended that you ask someone suitably qualified to do the jump starting for you.*

Before attempting to use jump leads, there are a few important points to note.

First of all, use only proper jump leads which have been specifically designed for the job.

Make sure that the booster battery (the fully-charged battery) is a 12-volt type.

Position the two vehicles close enough together to connect the leads, but **do not** allow the vehicles to touch.

Turn off all electrical circuits. Position the gear lever in neutral.

DO NOT allow the ends of the two jump leads to touch at any time during the following procedure.

Proceed as follows.

1 Open the vehicle bonnets, and connect one end of one of the jump leads (usually the red one) to the positive (+) terminal of the flat battery, and the other end to the positive terminal of the booster battery.

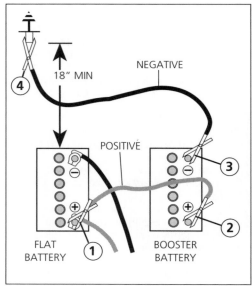

▲ *Jump start lead connections for negative earth vehicles – connect leads in order shown*

BREAKDOWNS — 49

2 Connect one end of the other jump lead to the negative (-) terminal of the booster battery, and connect the other end to a suitable earth point on the car with the flat battery at least 45 cm (18 in) away from the battery (eg a clean, bare metal area on the engine block or body). Make sure that all the lead clips are secure.

3 Start the engine of the vehicle with the booster battery and let it run for a few minutes, then start the engine of the car with the flat battery in the normal way. Depress the clutch pedal during starting to reduce drag from the gearbox.

4 When the engine is running smoothly, disconnect the jump leads in exactly the reverse order to that in which they were connected.

What to carry in case of a breakdown

Carrying the following items may help to reduce the inconvenience and annoyance caused if you're unlucky enough to break down during a journey:

- Alternator drivebelt (refer to *'Servicing'* on page 98 for details of renewal)
- Spare fuses (refer to *'Bulb, fuse and relay renewal'* on page 118 for details of renewal)
- Set of principal light bulbs (refer to *'Bulb, fuse and relay renewal'* on page 113 for details of renewal)
- Battery jump leads (refer to *'Starting a car with a flat battery'* on page 48 for details of starting using jump leads)
- Tow-rope (refer to *'Towing'* on page 46 for details of towing procedure)
- Litre of engine oil (refer to *'Regular checks'* on page 74 for details of checking oil level)
- Torch

50 DRIVING SAFETY

VAUXHALL NOVA

DRIVING SAFETY

In the UK, more people are injured or killed every year due to road accidents than through any other single cause.

Car manufacturers are paying more attention to passive safety when designing new cars, but drivers must ultimately take active responsibility of ensuring the safety of themselves, their passengers, and just as importantly, other road users.

There will always be accidents on the roads, but taking the time to read the following advice may help you to avoid such a mishap.

The following Sections aim to give advice which will help you to reduce the risk of becoming involved in an incident when driving, not necessarily by changing the way that you drive, but simply by making you aware of some of the potential risks which can easily be avoided. For more information and practical advice on road safety, it is well worth considering taking part in one of the various courses provided by organisations such as RoSPA and the Institute of Advanced Motorists.

BEFORE STARTING A JOURNEY

As a driver, before beginning a journey it's important to make sure that you're comfortable so that you can concentrate on driving without any unnecessary distractions. Spending a few moments carrying out a few simple checks and adjustments will help to make your journey more relaxing, safer and hopefully trouble-free.

Driver comfort

Before driving, make sure that you're comfortable, and that you can operate all the controls easily, particularly if someone else has recently driven the car.

- **Seat**
  Make sure that the seat is positioned a comfortable distance from the steering wheel and the pedals, so that you can operate all the controls comfortably
- **Seat back**
  Make sure that the seat back is adjusted to give your back plenty of support. Your back should rest against the seat, and there should be no need to lean forwards from the seat back. You should be able to reach the steering wheel comfortably, and you should be able to turn the wheel easily without stretching.
- **Head restraint**
  If a head restraint is fitted, it should be adjusted to support your head in the event of an accident. Your head should not rest against the restraint during normal driving and, as a rough guide, the restraint is correctly positioned when its top is in line with your eyes. Similarly, make sure that any passenger head restraints are adjusted correctly if passengers are being carried.
- **Mirrors**
  All the mirrors should be adjusted so that they give a clear view behind the car without the need to move your head unnecessarily.
- **Steering column**
  If the car is fitted with an adjustable steering column, adjust the column so that the steering wheel can be reached comfortably, and the wheel can be turned easily without stretching.

Checks

If you're planning a long journey, refer to *'Regular checks'* on page 73, and make sure that you carry out all the checks mentioned before setting off. Also make sure that you know where the spare wheel and the tools required for wheel changing are located (refer to *'Breakdowns'* on page 44).

VAUXHALL NOVA

DRIVING SAFETY

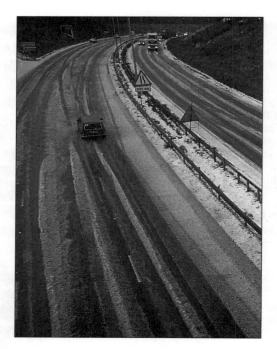

DRIVING IN BAD WEATHER

When driving in bad weather, always be prepared. Bad weather should never take you by surprise. Listen to a weather forecast: you should always be aware of the possibility of poor conditions on your intended journey.

Driving in bad weather requires more concentration, and is more tiring than driving in good weather conditions. Never allow yourself to be distracted by talkative passengers or loud music in the car and, if you feel tired, stop at the next opportunity and take a break.

Always look well ahead, so that you're aware of the condition of the road surface, and any obstacles; **slow down if necessary**.

The following advice is intended to help you to drive more safely in various bad weather conditions. However, above all, use common sense.

Rain

- Use dipped headlights in poor visibility.
- Slow down if visibility is poor or if there is a lot of water on the road surface.
- Keep a safe distance from the vehicle in front (stopping distances are doubled on a wet road surface).
- Be particularly careful after a long period of dry weather. Under these conditions, rain can make the road surface very slippery.
- Don't use rear foglights unless visibility is **seriously** reduced (generally less than 100 metres/300 feet). Foglights can dazzle drivers following behind, especially in motorway spray.

Fog

- Slow down. Fog is deceptive, and you may be driving faster than you think. Fog can also be patchy, and the visibility may suddenly be reduced. Always drive at a speed which allows you to stop in the distance you can see ahead.
- Use dipped headlights. Using main-beam headlights will usually reduce the visibility even further, as the fog will scatter the light.
- Use foglights (where they are fitted) if it's genuinely foggy (generally, where visibility is reduced to less than 100 metres/300 feet), and not just misty.
- Keep a safe distance from the vehicle in front.
- Use your windscreen wipers to clear moisture from the windscreen.

Snow

- Don't start a journey if there is any possibility that conditions may prevent you from reaching your destination. Listen to a weather forecast before setting off.
- Before starting a journey, clear **all** snow from the windscreen, windows and mirrors. Don't just clear a small area big enough to see through.
- Slow down.
- Keep a safe distance from the vehicle in front (stopping distances can be trebled or even quadrupled on a snow-covered or icy surface).
- Drive smoothly and gently. Accelerate gently, steer gently and brake gently.
- Don't brake and steer at the same time – this may cause a skid (refer to *'Skid control'* on page 56).
- When moving away from a standstill, or

VAUXHALL NOVA

DRIVING SAFETY

manoeuvring at junctions, etc, in a car with a manual gearbox use the highest possible gear that your car will accept to move away, and change up to a higher gear earlier than usual. This will provide more grip and will help to reduce wheelspin.
- Use main roads and motorways where possible. Major roads are likely to have been 'gritted' and are usually cleared before minor roads.
- Use dipped headlights in poor visibility.
- Don't use rear foglights unless visibility is **seriously** reduced (generally less than 100 metres/300 feet) – foglights can dazzle drivers following behind.

Ice and frost
- Follow the advice given for driving in snow.
- Be prepared for 'black ice'. Although you can't see 'black ice', reduced road noise from the tyres will usually tell you it's there.

Severe winter weather
The best advice in severe winter weather is to stay at home but, if you must drive your car, in addition to following the advice given for driving on snow and ice, there are a few additional pieces of advice which you should bear in mind before setting out on a journey.

- Make sure that you tell someone where you're going, and tell them roughly what time you're expecting to arrive at your destination, and what route you're taking.
- Keep a can of de-icer fluid, a scraper, a set of jump leads and a tow rope in the car at all times.
- Carry plenty of warm clothes and blankets.
- Make sure that you have a full tank of petrol before starting your journey. This will allow you to keep the engine running, without fear of running out of petrol, to provide warmth through the car's heating system should you be delayed by conditions, or worse still stuck.
- Pack some pieces of old sacking or similar material, which you can place under the driving wheels to give better traction if you get stuck.
- Carry a shovel, in case you need to dig yourself or someone else out of trouble.

MOTORWAY DRIVING

Motorway driving requires a great deal of concentration and awareness. Motorways are generally busier than ordinary roads, and the traffic moves faster, so there is less time to react to any changes in road conditions ahead.

This Section covers the fundamental rules for safe motorway driving, and will help you to avoid many of the problems which occur on today's busy motorways. Remember that lack of common sense is probably the biggest cause of accidents on motorways.

Full rules and regulations for motorway driving can be found in 'The Highway Code'.

Joining and leaving a motorway
When you join a motorway, you will normally approach from a 'slip-road' on the left. You must give way to traffic already on the motorway. Watch for a safe gap in the traffic, and adjust your speed in the acceleration lane so that when you join the left-hand lane of the motorway you're already travelling at the same speed as the other traffic. Indicate before pulling onto the motorway. After joining the motorway, allow yourself some time to get used to the speed of the traffic before overtaking.

When you leave a motorway, indicate in plenty of time, and reduce your speed before you enter the slip-road – some slip-roads and

VAUXHALL NOVA

link-roads between motorways have sharp bends which can only be taken safely by slowing down. Be very cautious immediately after leaving a motorway, as it can be difficult to judge speed after a long period of fast driving.

Rules for safe motorway driving

- **Concentrate and think ahead** – Always be prepared for the unexpected, and look well ahead. You should be aware of all the traffic in front and behind, not just the vehicle in front of you. If you're concentrating properly, nothing should take you by surprise on a motorway.
- **Always drive at a speed to suit the road conditions** – Don't break the speed limit, especially temporary speed limits which may apply to contra-flow systems or roadworks.
- **Slow down** in bad weather conditions. Driving too fast in fog and motorway spray is a major cause of motorway accidents.
- **Always keep a safe distance from the vehicle in front** – The closer you drive to the vehicle in front, the less chance you have of avoiding an accident if something happens ahead. Remember that you need more space in bad weather conditions.
- **Think** – If the vehicle ahead suddenly stops, do you have enough space to stop without hitting it?
- **Use your mirrors regularly** – It's just as important to be aware of what's happening behind you as it is in front.
- **Always signal your intentions clearly and in plenty of time** – Check your mirrors first, signal *before* you move, and change lanes smoothly and in plenty of time. Don't make any sudden moves which are likely to affect traffic approaching from behind.
- **Keep to the left** – The left-hand lane is *not* the slow lane, and you should drive in this lane whenever possible. You can stay in the middle lane when there are slower vehicles in the left-hand lane, but return to the left-hand lane when you've passed them. The right-hand lane is for overtaking only: it is *not* for driving at a constant high speed. If you use the right-hand lane, move back into the middle lane and then into the left-hand lane as soon as possible, but without cutting in.

- **Take note of direction signs** – You may need to change lanes to follow a certain route, in which case you should do so in plenty of time.
- **Stop as soon as practicable if you feel tired** – Driving on a motorway when feeling tired can be extremely dangerous. If necessary, wind down the window for fresh air. If you feel tired or drowsy, stop at the next service station, or turn off the motorway at the next junction and take a break before continuing your journey. You must not stop on the hard shoulder of a motorway other than in an emergency.
- **If you break down** – Refer to 'Breakdowns' on page 43.

TOWING A TRAILER OR CARAVAN

When towing a trailer or a caravan, there are a few special points to bear in mind. There's more to towing a trailer or caravan than just simply bolting on a towbar and hitching up!

The law

Before using your car for towing, make sure that you're familiar with any special legislation which may apply, particularly if you're travelling abroad.

Make sure that you know the speed limits applicable for towing.

In some countries, including the United Kingdom, there is a legal requirement to have

DRIVING SAFETY

a separate warning light fitted in the car to show that the trailer/caravan direction indicator lights are working.

Always check on current regulations before you travel.

Before starting a journey

Before attempting to use your car for towing, there are one or two points which should be considered.

- **Make sure that your car can cope with the load which you're towing**
- **Don't exceed the maximum trailer or towbar weights for the car** – Refer to *'Dimensions and weights'* on page 13.
- **Engine** – Don't put unnecessary strain on your car's engine by trying to tow a very heavy load. Obviously, the smaller the engine, the less load the car can comfortably tow. Also bear in mind that the extra load on the engine when towing means that the engine's cooling system may no longer be adequate. Some manufacturers provide modified cooling system components for towing, such as larger radiators, etc, and if you intend to use your car for towing regularly, it may be worth enquiring about the availability and fitting of these components.
- **Suspension** – Towing puts extra strain on the suspension components, and also affects the handling of the car, since standard suspension components aren't usually designed to cope with towing. Heavy duty rear suspension components are available for most cars, and it's worth considering having these fitted if you intend to use your car for towing regularly. Special stabilisers are also available for fitting to most cars to reduce pitching and 'snaking' movements when towing.
- **Make sure that you can see behind the trailer/caravan using the car's mirrors** – Additional side mirrors with extended arms are available to fit most cars. The mirrors must be fitted with folding arms (so that they fold if hit), and should be adjusted to give a good view to the rear at all times. The mirrors can be removed when the car isn't being used for towing.
- **Make sure that the tyre pressures are correct** – Unless a light, unladen trailer is being towed, the car tyres should be inflated to their 'full load' pressures. Also make sure that where applicable the trailer or caravan tyres are inflated to their recommended pressure.
- **Make sure that the headlights are set correctly** – Check the headlight aim with the trailer/caravan attached, and if necessary adjust the settings to avoid dazzling other drivers.
- **Make sure that the trailer/caravan lights work correctly** – Make sure that the tail lights, brake lights, direction indicator lights and, where applicable, the reversing lights and fog lights all work correctly. Note that the additional load on the direction indicator circuit may cause the lights to flash at a rate slower than the legal limit, and in this case you may need to fit a 'heavy duty' flasher unit.
- **Make sure that the trailer/caravan is correctly loaded** – Refer to the manufacturer's recommendations for details of loading. As a general rule, distribute the weight so that the heaviest items are as near as possible to the trailer/caravan axle. Secure all heavy items so that they can't move. To provide the best possible control and handling of the car when towing, the manufacturers recommend an optimum noseweight for the trailer/caravan when loaded (refer to *'Dimensions and weights'* on page 13 for details). The noseweight can be measured using a set of bathroom scales as follows:

Place a stout piece of wood between the trailer/caravan tow hitch cup and the scales platform, and read off the weight with the trailer/caravan level. If necessary, redistribute the load, to arrive as close as possible to the recommended noseweight. **Do not** exceed the maximum recommended noseweight.

Driving tips

Towing a trailer or caravan will obviously affect the handling of the car, and the following advice should be followed when towing.

- **If possible, avoid driving with an unladen car and a loaded trailer/caravan** – The uneven weight distribution will tend to make the car unstable. If this is unavoidable, drive slowly to allow for the instability.

DRIVING SAFETY

- **Always drive at a safe speed** – The stability of the car and the trailer or caravan decreases as the speed increases. Always drive at a speed which suits the road and weather conditions, and always reduce speed in bad weather and high winds – especially when driving downhill. If the trailer or caravan shows any sign of 'snaking', reduce speed immediately – never try to stop 'snaking' by accelerating.
- **Always brake in good time** – If you're towing a trailer or caravan which has brakes, apply the brakes gently at first, then brake firmly. This will help to prevent the trailer wheels from locking. In cars with a manual gearbox, change into a lower gear before going down a steep hill so that the engine can act as a brake, and similarly, in cars with automatic transmission, move the selector lever to position '2', or '1' in the case of very steep hills.
- **Don't change to a lower gear unnecessarily** – Unless the engine is labouring, stay in as high a gear as possible to keep the engine revs as low as possible. This will help to avoid the engine overheating.

ALCOHOL AND DRIVING

By far the best advice on drinking and driving is **DON'T**.

You may feel fine, but it's a proven fact that even one small alcoholic drink will impair your driving to some extent.

Drinking alcohol has the following effects:

- **Reduces co-ordination**
- **Increases reaction time**
- **Impairs judgement of speed, distance and risk**
- **Encourages a false sense of confidence**

The risk of an accident increases sharply after drinking alcohol, and approximately one third of the total number of people killed in road accidents each year have blood alcohol levels above the legal limit for driving. Remember that you're putting other people as well as yourself at risk if you drive after drinking.

The legal limit for blood alcohol level in the UK is 80 milligrams per 100 millilitres. This doesn't correspond to any particular quantity of drink, as the amount of drink required to reach this level varies from person to person. The driving of many people who feel perfectly sober is seriously affected well below the legal limit.

The penalties for driving over the legal limit are severe, and can mean losing your driving licence, a heavy fine, or imprisonment. It's also important to realise that the laws abroad can be far more severe than in the UK, and some countries have a total ban on driving after drinking alcohol.

- **The safest course of action is not to drink and drive.**

SKID CONTROL

If you drive sensibly with due regard for the road conditions, you should never find yourself in a situation where your car is skidding. The following advice will help you to avoid situations which may cause a car to skid, and explains how to regain control of your car quickly should the need arise.

VAUXHALL NOVA

DRIVING SAFETY

A skid is caused by one or a combination of the following:

- **Excessive speed in relation to the road conditions.**
- **Harsh or excessive acceleration.**
- **Sudden or excessive braking.**
- **Coarse or excessive steering.**

The most common basic cause of skidding is rough handling of the car's controls. Therefore the key to safe driving and preventing a skid is smooth, gentle handling of the car and its controls. Try to apply smooth pressure to the brake pedal, accelerator and steering wheel rather than just suddenly moving them a certain distance.

When a car is skidding, the following things happen:

- **The car is out of control. You can take action to regain control, but while skidding, the car can't be fully controlled.**
- **A car with the front wheels skidding can't be steered.**
- **A car with the rear wheels skidding is likely to spin round if any steering lock is applied.**
- **A car with all four wheels skidding will continue in a straight line in the direction it was travelling when the skid started, regardless of which way the car is pointing.**

Prevention is better than cure. When approaching a corner, reduce speed early by smooth progressive braking **in a straight line**. In a car with a manual gearbox, brake to an appropriate speed and select the correct gear for the corner. Turn into the corner early and smoothly, gently increasing the steering lock if necessary, and corner with your foot gently on the accelerator, so that the engine just 'pulls' the car through the corner, maintaining a constant speed. As you leave the corner, accelerate gently and smoothly.

If you follow the above advice, you should avoid finding yourself in a situation where your car is skidding. However, should you be unfortunate enough to experience skidding, the basic rule for skid control is to **remove the cause** of the skid. If you cause a skid by accelerating, braking or steering, **ease off**, and, when the skid stops, re-apply the accelerator, brake or steering control, but this time more gently and smoothly. If the front wheels are skidding, to regain steering control quickly, steer 'into the skid'. This is often misunderstood, and it basically means look in the direction you want to be facing, and steer in that direction – and be prepared to reduce the steering quickly if necessary when the wheels regain their grip.

Several organisations offer courses in skid control, using specially-equipped cars under carefully-controlled conditions. One of these courses could prove to be a very worthwhile investment, possibly helping you to avoid an accident.

ADVICE TO WOMEN DRIVERS

Unfortunately, it's a fact that women are more likely than men to be the subject of unwelcome attention when driving alone.

There's no reason to think that driving alone spells trouble, but it's a good idea to be aware of certain precautions which will help to avoid finding yourself in an unpleasant situation.

This Section aims to give advice which will help to reduce any risks when driving alone, and will help you to deal with the situation, should you be unfortunate enough to find yourself the subject of unwelcome attention.

Several organisations (some local police authorities, AA, etc) run courses especially for women drivers, in which subjects such as basic car maintenance and self defence, etc, are covered.

Sensible precautions
- **The car**

Make sure that you're familiar with your car.

Make sure that your car is regularly serviced. Refer to *'Regular checks'* on page 73, and make sure that all the checks described are carried out regularly. This will minimise the possibility of a breakdown.

Make sure that you know how to change a wheel, and make sure that the car jack is in good condition in case you have a puncture - refer to *'Breakdowns'* on page 44.

VAUXHALL NOVA

DRIVING SAFETY

● The journey

If you're planning to undertake a particularly long or unfamiliar journey, the following advice may be helpful.

If you're travelling on an unfamiliar route, make a few notes before you set off, reminding yourself of the road numbers, where to turn, which junctions to use on motorways, etc. Try to stick to main roads where possible. Always carry a map.

Make sure that you have enough petrol for the journey. If you need to fill up during the journey, make sure that you do so in plenty of time, before the fuel level gets too low, and before petrol stations close if travelling at night.

If you think that your family or friends may be worried, you may want to 'phone someone at your destination before setting off to tell them that you're leaving, and tell them what time you expect to arrive. Similarly, when you arrive, you may want to 'phone someone at your starting point to confirm that you've arrived safely.

● Driving in traffic

Avoid attracting unnecessary attention.

Avoid 'jokey' stickers in car windows, which may encourage unwelcome attention.

Avoid eye contact with 'undesirables'.

If you're being followed, pull over and slow down, but **don't** stop. **Don't** drive to your home, but find a busy and well lit place. If the person following persists, blow your horn and flash your lights to attract attention.

If you're stopped by traffic or another vehicle, lock the doors and close the windows. **Don't** ram the other vehicle - damage to your car might prevent your escape.

● Leaving your car

Don't leave any clues that the car is being driven by a woman; make sure that any 'feminine' items are hidden from sight before leaving your car.

Refer to *'Car crime prevention'* on page 67 and take note of the advice given.

Try to park in a well-lit, preferably busy area.

If you park in a car park, try to park close to an exit, or close to the attendant's station.

Always reverse into a parking space, so that you can drive away quickly if necessary after returning to your car.

Take note of any 'landmarks' so that you can find your car quickly when you return.

Always lock your car.

When you return to your car, walk with a group of people, if possible. Have the keys ready so that you don't have to spend unnecessary time outside your car searching for them, which may attract attention.

Before getting into your car, briefly check for forced entry, and look into the car for any suspicious signs. **Don't** get into the car if you notice anything suspicious.

● Breakdowns

Remember that you're more likely to be injured in an accident than a personal attack.

Refer to *'Breakdowns'* on page 43 and take note of the advice given but, in addition, bear in mind the following advice.

Always walk to 'phone for assistance, **don't** accept a lift. When you 'phone for assistance, mention that you're an unaccompanied woman.

If someone stops while you're 'phoning for help, give the operator details of the other vehicle's registration number and a brief description of the car and driver. If the driver approaches you, tell them that you have passed on his/her details to the police. If the driver's intentions are honourable, your reaction will be understood.

If you decide to stay with, but outside the car, leave the nearside door unlocked and slightly open, so that you can get inside quickly to lock yourself in if necessary.

If you decide to stay inside the car, sit in the passenger seat and lock the door - this will give the appearance that you're accompanied.

When help arrives, ask for some form of identification (even from a policeman) before giving any of your own details.

If someone offers assistance, tell them the police have been informed and are arranging recovery. If you have not yet contacted the police, consider asking the person to do so on your behalf, **but** if you're uncertain about the person's intentions, tell them that the police are aware and ask them to call the police again for you.

DRIVING SAFETY 59

● **Accidents**

Keep calm.

Refer to 'Accidents and emergencies' on page 35 and take note of the advice given but, in addition, bear in mind the following advice.

If you're bullied or shouted at, lock yourself in your car (if it's safe to do so), and wind the windows up. Communicate through a small gap at the top of the window.

If things get out of hand, refuse to talk to anyone except the police.

Self defence

This is a last resort and should not be used unless all else has failed.

There is no substitute for taking a properly supervised course in self-defence, but we aim to outline a few of the basic moves below.

If you are approached, act in a composed manner and use a soft tone.

In the event of an attack, stay calm - you can still safeguard yourself.

- **If you're held by one arm** - Use your free hand to grasp the attacker's thumb and twist the thumb sharply back towards the wrist.
- **If you're facing your attacker** - If your shoulders are held, thrust your hands/arms upwards and diagonally outwards to dislodge the attacker's grip.

    Use your knees or the blade of your hand to chop your attacker's groin, your feet to kick the shins, and your fingers to poke the eyes.
- **If you're attacked from behind** - Quickly lean forward, twisting your head sideways into the attacker to keep your airway (for breathing) clear. Often this will cause the attacker to lose balance.
- **If you've been pulled backwards already** - Turn your head as above, and chop hard with the blade of your hand or a clenched fist to the groin, or an elbow in the stomach.
- **If you've been forced to the floor** - Use your feet and legs to kick against the attacker's shins whilst swivelling your body to keep the attacker at bay.

Be ready to run as soon as the attacker's grip is released, and SCREAM!

CHILD SAFETY

Every day, thousands of parents strap their offspring into child car seats, confident that they have done everything possible to protect their loved ones. However, a recent survey has shown that many young children are travelling in car seats which have been incorrectly fitted or are being incorrectly used, making them potentially dangerous should they be involved in an accident. The following advice will help you to make sure that you take every possible precaution to ensure the safety of your children when driving.

How to avoid unnecessary risks

- **Never allow young children to travel in a car unrestrained,** even for the shortest of journeys.
- **Never carry a child on an adult's lap or in an adult's arms.** Although you may feel that your baby is safer in your arms, this is not the case. An adult holding a child is far more likely to cause injury to the child than to give protection in the event of an accident.

VAUXHALL NOVA

DRIVING SAFETY

- **Never rely solely on an adult seat belt to restrain a child,** and never sit a child on a cushion to enable a seat belt to fit properly.
- **Always strap young children into a properly designed child car seat** when carrying them in a car. The cost of a good car seat is a very small price to pay to save your child's life.

Choosing a child car seat

- It's vitally important to choose the correct type of car seat for your child. A wide range of child car seats is available, and it's worth spending some time looking at the various seats on the market before deciding on the most suitable seat for your particular requirements.
- Although age ranges are often given by the manufacturers, these should be taken as a rough guide. It's the weight of the child which is important; for example, a smaller than average baby could use a baby car seat for longer than a heavy baby of the same age.
- **Never** buy a secondhand child seat. This may sound like a ploy from the manufacturers to sell more seats, but all too often a secondhand seat is sold without the instructions, and sometimes there are parts missing. This often leads to secondhand seats being incorrectly fitted. A secondhand seat may have been damaged or weakened through carelessness or misuse, without necessarily showing any visible signs until it's put to the test in an accident!
- Your baby's first contact with the outside world is often on that first ride home from hospital, so make sure that you're prepared, and buy a car seat before your baby is born.
- Some seats are designed for babies from birth to a weight of around 22 pounds (10 kg), which for most babies is around nine months old. Usually, these first car seats are light and easy to transport through the use of a handle. This means that a sleeping baby can be carried from the car into the house without waking. Some baby seats come complete with a built-in headrest, which is very important for the early weeks when a baby's head needs to be supported.

Certain seats can be fitted so that the baby faces the back of the car (ie rearward facing seats). This may be considered to improve safety, as the baby is supported across the back rather than purely by the harness, if the car is involved in a frontal impact. Combination type seats can then be used forward facing when the child reaches approximately nine months old.

Other types of seat may be designed to accommodate children up to around 40 pounds (18 kg) or approximately four years old.

Alternatively, some seats use the car's seat belts to hold both the child seat and the child, and these seats are obviously easier to fit. Make sure that this type of seat is fitted with a seat-belt lock, so that the seat belt cannot be pulled out of place or slackened. Another advantage of these seats is that they can easily be transferred from car to car.

- Try to choose a seat which has an easily adjustable harness, as this makes it easier to ensure that the harness fits the child securely for each trip. If the harness can be quickly loosened, it makes getting a struggling child in and out a little easier.

Using a child car seat

- Firstly, make absolutely sure that the seat is properly fitted in accordance with the manufacturer's instructions. If you're unsure about any of the fitting procedure, contact the manufacturer for advice.
- To hold a child securely, the harness must be reasonably tight. There should be just enough room to slide your flat hand under the strap. Children may be wearing bulky clothes one day, and thin clothes the next, so it is vital that the harness is adjusted before each journey to ensure a correct fit.
- If the seat is fitted using an adult's inertia reel seat belt, make sure that the seat is held firmly in position, and that the seat belt is securely locked to prevent it from loosening. Also, make sure that the seat belt buckle is not resting on the frame of the child seat. This is because the buckles are not designed to withstand the impact of a heavy child seat and, in an accident, could break open.

DRIVING ABROAD

Driving abroad can be very different to driving in the UK. Besides the obvious differences, like climate and driving on the right-hand side of the road, various unfamiliar laws may apply, and it's advisable to prepare yourself and your car as far as possible before travelling.

This Section provides you with a guide which will help you to prepare for driving abroad, and will help you to avoid some of the pitfalls waiting for the unwary.

It may be worthwhile considering hiring a car abroad, rather than driving your own car. In this case, make sure that the insurance cover arranged suits your requirements, and make sure that a damage waiver is included (otherwise you will have to pay for any damage to the hire car).

Insurance
● Motoring
Make sure that you have adequate insurance cover for your car and your luggage.

Check on the legal requirements for insurance in the country you're visiting, and always inform your insurance company that you're taking your car abroad – they will be able to advise you of any special requirements.

Most car insurance policies automatically give the minimum legally-required cover for driving in EC countries, but if you require the same level of cover as you have in the UK, you will normally need to obtain a 'Green Card' (an internationally-recognised certificate of insurance) from your insurance company.

● Medical
It's always advisable to take out medical insurance for the car occupants. Not all countries have a free emergency medical service, and you could find yourself with a large unexpected bill in the event of yourself or one of your passengers being taken ill, or being involved in an accident (in some countries you may even have to pay for an ambulance).

NHS form E111 (available by filling in a form at a Post Office) entitles you to receive

DRIVING ABROAD

the same health care that residents receive in EC countries, but this is by no means comprehensive, and additional insurance cover is usually advisable.

● **Breakdown**

Recovery and breakdown costs can be far higher abroad than in the UK.

Most of the national motoring organisations (such as the AA and RAC) will be able to provide insurance cover which could save you a lot of inconvenience and expense should you be unfortunate enough to break down.

Documents

Always carry your passport, driving licence, car registration document, and insurance certificate (including 'Green Card' and medical insurance, where applicable).

Make sure that all the documents are valid, and that the car's road fund licence (tax) and MOT don't run out while you're abroad.

Before travelling, check with the authorities in the country you're visiting, in case any special documents or permits are required. You may need a visa to visit some countries, and an International Driving Permit (available from the AA or RAC) is sometimes required.

Driving laws

Before driving abroad, make sure that you're familiar with the driving laws in the country you're visiting, as there may be some laws which don't apply in the UK, and the penalties for breaking the law may be severe.

Fit a 'GB' plate to the back of your car, and make sure that it's displayed all the time you're abroad.

In some countries, you're legally required to carry certain items of safety equipment. These can include a first-aid kit, a warning triangle, a fire extinguisher, a set of spare bulbs/fuses, etc.

Remember that if you're visiting a country where you have to drive on the right-hand side of the road, you'll need to fit headlight beam deflectors or shields to avoid dazzling other drivers, or alternatively, it may be possible to have the headlight beams adjusted.

Make sure that you're familiar with the speed limits, and note that in some countries there is an absolute ban on driving after drinking *any* alcohol.

Servicing

Service your car before setting off on your trip, to reduce the possibility of any unexpected breakdowns.

Pay particular attention to the condition of the battery, windscreen wipers and tyres (including the spare), noting that the tyre pressures will probably have to be increased from their normal setting if the car is to be fully loaded. If you're travelling to a cold country, check the condition of the cooling system and the strength of the antifreeze.

Before setting off on your journey, refer to '*Regular checks*' on page 73, and carry out all the checks described. Check that the jack and wheel brace are in place in the car (refer to '*Breakdowns*' on page 44), and that the jack works properly.

Spares

In addition to the items which must be carried by law in the country you're visiting, it's a good idea to carry a few spares which may be difficult to obtain should you need them abroad (one of the national motoring organisations should be able to advise you). For example, you may want to carry clutch and throttle cable repair kits, as right-hand-drive components can be difficult to find outside the UK.

If your car uses a special oil, it's a good idea to take a pack with you.

It's also a good idea to carry a tow rope and a set of jump leads to help you out in case of a breakdown.

Fuel

The type and quality of petrol available varies from country to country, and it's a good idea to check on the availability of the correct petrol type before travelling (again, one of the national motoring organisations will be able to advise you). This is especially important if your car has a catalytic converter, as in this case you must only use unleaded petrol.

Find out what petrol pump markings to look for to give you the correct type and grade of petrol for your car.

Security

A foreign car packed with luggage is an inviting prospect for criminals, so refer to '*Car*

DRIVING ABROAD

crime prevention' on page 67, and don't take any chances.

Route planning

It's always a good idea to plan your approximate route before travelling. A vast number of maps and guides are available, or the AA and RAC can provide you with directions to your destination for a modest charge.

Bear in mind that in some countries, you'll have to pay tolls to use certain roads, and this can add unexpectedly to the cost of travelling.

What to carry when driving abroad

The following list provides a guide to the items which it is compulsory to use or carry, or it is strongly recommended that you carry in your car when driving in various European countries.

In the following table **'C'** indicates **'Compulsory'** and **'R'** indicates **'Recommended'**.

COUNTRY	HEADLAMP DEFLECTORS	GB STICKER	SEAT BELTS	WARNING TRIANGLE	FIRE EXTINGUISHER	FIRST AID KIT	SPARE BULBS
Austria	C	C	C	C	-	C	-
Belgium	C	C	C	C	C	-	-
Bulgaria	C	C	C	C	C	C	-
Czechoslovakia	C	C	C	C	-	C	C
Denmark	C	C	C	R	-	-	-
Eire	-	C	C	-	-	-	-
Finland	C	C	C	C	-	-	-
France	C	C	C	C	-	-	R
Germany	C	C	C	C	C	C	R
Greece	C	C	C	C	C	C	-
Holland	C	C	C	C	-	-	C
Hungary	C	C	C	C	-	-	C
Italy	C	C	C	C	-	-	R
Luxembourg	C	C	C	R	-	-	-
Norway	C	C	C	R	-	-	R
Poland	C	C	C	C	-	-	R
Portugal	C	C	C	C	C	-	-
Spain	C	C	C	C	-	-	C
Sweden	C	C	C	R	-	-	-
Switzerland	C	C	C	C	-	-	-
United Kingdom	-	-	C	-	-	-	-
Yugoslavia	C	C	C	C	-	C	C

VAUXHALL NOVA

64 REDUCING THE COST OF MOTORING

VAUXHALL NOVA

REDUCING THE COST OF MOTORING

ECONOMICAL DRIVING

Owning a car is always going to involve some expense, and the running costs can generally be divided into two main areas. The first concerns fixed costs which cannot be avoided, such as car tax and insurance (although obviously you can shop around for the best insurance quote). However, for most owners savings can be made in the second main area of expense, which covers fuel costs and servicing/maintenance bills.

It's surprisingly easy to reduce the amount of fuel used, simply by adapting your driving style to suit the prevailing conditions, and avoiding certain driving habits which tend to increase fuel consumption unnecessarily.

A large proportion of the average servicing bill is made up of garage labour time, money which can be saved by carrying out the work yourself. Refer to *'Servicing'* on page 81 for easy-to-follow instructions on how to carry out most servicing work. It's also worth bearing in mind that maintenance costs can be cut significantly by reducing unnecessary wear-and-tear on the car.

The following advice deals with the most significant causes of high fuel consumption and unnecessary wear-and-tear, and explains how to save money and reduce environmental pollution during everyday driving.

- **Don't warm the engine up with the car standing still** – Engine wear and pollution is at its highest when the engine is warming up, and the engine takes a long time to warm up when running at idle speed with the car stationary. To avoid excessive wear and pollution, drive off as soon as the engine starts, and don't use more 'revs' than necessary.

- **On cars with a manual choke, don't use the choke any longer than necessary** – Push the choke control fully in as soon as the engine will run smoothly without it. When the engine is running with the choke applied, extra fuel is being used and so the fuel consumption is increased, and more pollution is produced.

- **Avoid sudden full throttle acceleration** – Sudden acceleration increases fuel consumption, engine wear and pollution.

- **Don't drive at high engine speeds** – Minimum fuel consumption and pollution is achieved at low engine speeds and in the highest possible gear. Lower engine speed also means less noise and engine wear. For maximum economy, stay in as high a gear as possible for as long as possible, without making the engine labour.

- **Don't always drive at maximum speed** – Fuel consumption, pollution and noise increase rapidly at high speeds. A small reduction in speed (particularly during motorway driving) can significantly lower fuel consumption and pollution.

- **Look well ahead, and drive as smoothly as possible** – By looking well ahead you will be able to react to any change in road conditions in plenty of time, allowing you to brake and accelerate smoothly. Unnecessary or harsh acceleration and braking increases fuel consumption and pollution.

- **Where possible, avoid dense, slow-moving traffic** – In these conditions, more frequent braking, acceleration and gear changing are required, which increase fuel consumption and pollution.

- **Stop the engine during traffic hold-ups** – Obviously if your engine has stopped it doesn't use any fuel and there is no pollution.

VAUXHALL NOVA

REDUCING THE COST OF MOTORING

- **Check the tyre pressures regularly** – Low tyre pressures increase the rolling resistance of the car, and therefore increase fuel consumption, as well as increasing tyre wear and causing handling problems.

- **Don't carry unnecessary luggage** – Weight has a significant effect on fuel consumption, especially in dense traffic where frequent acceleration is required.

- **Don't leave a roof rack fitted when not in use** – The extra air resistance increases fuel consumption.

- **Switch off any unnecessary electrical circuits as soon as possible** – Heated rear windows, foglights, heater blowers, etc, consume a considerable amount of electrical power. The engine must work harder to provide this power by driving the alternator, and the fuel consumption is therefore increased.

- **Check the fuel consumption regularly** – By doing this you will be able to notice any significant increase in fuel consumption, and any problem which may be causing it can be investigated before it develops into anything more serious.

- **Ensure that your car is serviced regularly** – This will ensure that the car operates as efficiently as possible, reducing fuel consumption and pollution.

- **If your car engine is suitable, use unleaded petrol** – This will reduce pollution. If in any doubt as to whether your car's engine is suitable for use with unleaded petrol, seek advice from the car's manufacturer, or from a recognised dealer.

VAUXHALL NOVA

CAR CRIME PREVENTION

Crime against cars is a serious problem, and many owners suffer the consequences of theft or damage every year. The likelihood of your car becoming a target for criminals can be reduced by making it more difficult for your car to be broken into or stolen, and the following advice should prove helpful. Some of the points may seem obvious, but the majority of cars are broken into or stolen in a very short space of time with little force or effort required.

- **Always remove the ignition key** – even in a garage or driveway at home. Always make sure that the steering column lock is engaged after removing the key.
- **Always lock your car** – even in a garage or driveway at home. Where fitted, ensure that 'deadlocks' are engaged before leaving the car and, where applicable, don't forget to lock the fuel filler cover. Make sure that all the windows and the sunroof or folding roof, where applicable, are properly shut. If the car is in a garage, lock the garage. If your car is stolen or broken into while it's unlocked, your insurance company may not pay for the full value of loss or damage.

- **Never leave valuable items on display** – Even if you're only leaving your car for a few minutes, move anything that might be attractive to a thief (even a coat or briefcase) out of sight, and preferably take it with you or lock it safely in the boot. Don't leave valuable items (especially credit cards) in the glovebox. Don't leave your vehicle documents in the car (registration document, MOT certificate, insurance certificate, etc), as they could help a thief to sell it.
- **Park in a visible and (preferably) busy area** – This will deter thieves as they run a greater risk of being caught. If you're parking your car at night, try to leave it in a well-lit area.
- **Put your radio aerial down (where applicable) when parking** – Radio aerials can prove an attractive target for vandals.
- **Protect any in-car entertainment equipment** – The latest security-coded equipment won't work if someone tampers with it and disconnects it from the battery. Some equipment is specially designed so that you can remove the control panel or the entire unit, and take it with you when you leave the car.
- **Fit lockable wheel nuts (or bolts, as applicable)** – if your car is fitted with expensive alloy wheels. Alloy wheels are a favourite target for thieves.
- **Have your car windows etched with the registration number** – This will help to trace your car if it's stolen and the thieves try to change its identity. Other glass components such as sunroofs and headlamps can also be etched if desired. Many garages and specialists can provide this service, and it's possible to buy DIY glass etching kits from motor accessory shops if you prefer to tackle the job yourself.
- **Fit a vehicle immobiliser device** – Many different types are available, but the most common types consist of a substantial metal bar which can be locked in place between the steering wheel and the pedals to prevent the car from being driven.
- **Have an alarm fitted** – Many different types are available, and some are expensive, but they will deter thieves. Some alarms have built-in immobiliser devices to prevent the car from being driven. If you have an alarm fitted, remember to switch it on even if you're only leaving the car for a few minutes.

VAUXHALL NOVA

SERVICE SPECIFICATIONS

VAUXHALL NOVA

SERVICE SPECIFICATIONS

Recommended lubricants and fluids

Component/system	Lubricant/fluid type and specification	Duckhams recommendation
ENGINE (1)	Multigrade engine oil, viscosity range SAE 10W/40 to 20W/50, to API SF or SG	Duckhams QXR, QS, Hypergrade Plus or Hypergrade
COOLING SYSTEM (2)	55% clean water and 45% ethylene glycol-based antifreeze	Duckhams Universal Antifreeze and Summer Coolant
BRAKING SYSTEM (3)	Hydraulic brake fluid to SAE J1703 or DOT 4	Duckhams Universal Brake and Clutch Fluid
MANUAL GEARBOX (4)	Gear oil, viscosity SAE 80, GM part number 90 001 777 or 90 188 629	Duckhams Hypoid 80

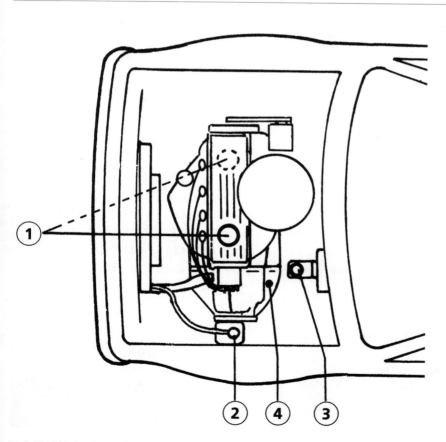

VAUXHALL NOVA

SERVICE SPECIFICATIONS

Lubricant and fluid capacities [Litres (Pints)]

Fuel tank capacity
All models	42 litres (9.3 gallons)

Engine oil
Quantity of oil required to bring level on dipstick from 'MIN' to 'MAX' mark:
1.0, 1.2, 1.3 & 1.4 litre engines	**0.75** (1.3)
1.6 litre engine	**1.0** (1.8)

Engine oil capacity (including filter)
1.0 litre engine	**2.5** (4.4)
1.2, 1.3 & 1.4 litre engines	**3.0** (5.2)
1.6 litre engine	**3.5** (6.2)

Coolant capacity (for coolant change)
1.0 litre engine	**5.5** (9.7)
1.2, 1.3 & 1.4 litre engines	**6.3** (11.1)
1.6 litre engine	**6.1** (10.7)

Contact breaker points
Points gap [mm (in)]	**0.40** (0.016)

Spark plugs

Type	AC Delco	Champion
1.0 litre engine up to August 1986	**R42 6FS**	**RL82YCC** or **RL82YC**
1.2, 1.3 & 1.4 litre engines up to August 1986	**R42 XLS**	**RN7YCC** or **RN7YC**
1.0 litre engine from August 1986	**CR42 CFS**	**RL82YCC** or **RL82YC**
1.2, 1.3, 1.4 & 1.6 litre engines from August 1986	**CR42 CXLS**	**RN7YCC** or **RN7YC**

Electrode gap [mm (in)]
AC Delco (all types)	**0.7** to **0.8** (0.028 to 0.032)
Champion RL82YCC, RN7YCC	**0.8** (0.032)
Champion RL82YC, RN7YC	**0.7** (0.028)

VAUXHALL NOVA

SERVICE SPECIFICATIONS

Clutch [mm (in)]

Clutch pedal stroke   **124.0** to **131.0** (4.9 to 5.2)

Tyre pressures – cold [bars (lbf/in^2)]

Note: *The following is intended as a guide only. Manufacturers frequently change tyre pressure recommendations, and it is suggested that a Vauxhall dealer is consulted for latest recommendations.*

Up to 3 occupants	Front	Rear
135 SR 13	**1.9** (28)	**1.7** (25)
145 SR 13 (models up to September 1987)	**1.6** (23)	**1.6** (23)
145 SR 13 and 145 TR 13 (models from September 1987)	**1.7** (25)	**1.7** (25)
155/70 SR 13 and 165/65 SR 14	**1.7** (25)	**1.7** (25)
165/65 TR 14	**1.9** (28)	**1.7** (25)
165/70 TR 13	**1.7** (25)	**1.7** (25)
175/65 HR 14	**1.8** (26)	**1.6** (23)
175/70 SR 13 (models up to September 1987)	**1.5** (22)	**1.5** (22)
175/70 SR 13 (models from September 1987)	**1.7** (25)	**1.7** (25)
Fully laden		
135 SR 13	**2.1** (31)	**2.6** (38)
145 SR 13 (models up to September 1987)	**1.8** (26)	**2.4** (35)
145 SR 13 and 145 TR 13 (models from September 1987)	**2.0** (29)	**2.4** (35)
155/70 SR 13 and 165/65 SR 14	**1.9** (28)	**2.4** (35)
165/65 TR 14 and 165/70 TR 13	**2.0** (29)	**2.4** (35)
175/65 HR 14	**2.0** (29)	**2.2** (32)
175/70 SR 13 (models up to September 1987)	**1.7** (25)	**2.3** (33)
175/70 SR 13 (models from September 1987)	**2.0** (29)	**2.4** (35)

REGULAR CHECKS

VAUXHALL NOVA

REGULAR CHECKS

To ensure that your car is reliable and safe to drive, there are one or two essential checks which are so simple that they're often ignored. These checks only take a few minutes, and could save you a lot of inconvenience and expense. It's a good idea to carry out these checks once a week, and certainly before you start off on a long journey. This Section explains how to carry out the checks, and what to do if things aren't quite as they should be.

Whenever you're carrying out checks or servicing jobs, safety must always be the first consideration, and you should bear in mind the advice given in the *'Safety first!'* notes on page 86 before proceeding.

CHECKS

The following checks should be carried out regularly. The checks are explained in more detail in the following pages.

- *Check the oil level*
- *Check the coolant level*
- *Check the brake fluid level*
- *Check the tyres*
- *Check the washer fluid level*
- *Check the battery electrolyte level (where applicable)*
- *Check the wipers and washers*
- *Check the lights and horn*
- *Check for fluid leaks*

ITEMS REQUIRED WHEN CARRYING OUT CHECKS

When carrying out the regular checks, it's a good idea to have the following items close-to-hand. You won't always need all the items, but it's as well to have them available just in case.

- *Small quantity of clean rag* ● *1.0 litre pack of engine oil (of the correct type – refer to* **'Service specifications'** *on page 69)* ● *Small quantity of coolant solution (made up from approximately 55% clean water and 45% antifreeze)* ● *Small container of brake fluid (must be an airtight container)* ● *Tyre pressure gauge and foot pump* ● *Small screwdriver (for removing stones from tyres)* ● *Washer fluid additive* ● *Small quantity of distilled water – for models fitted with batteries which require topping-up* ● *Pin (for adjusting washer nozzles)*

There are several different engine sizes and types fitted to the Nova range. Refer to *'Servicing'* on page 89 for details of how to distinguish one engine from another.

VAUXHALL NOVA

74 REGULAR CHECKS

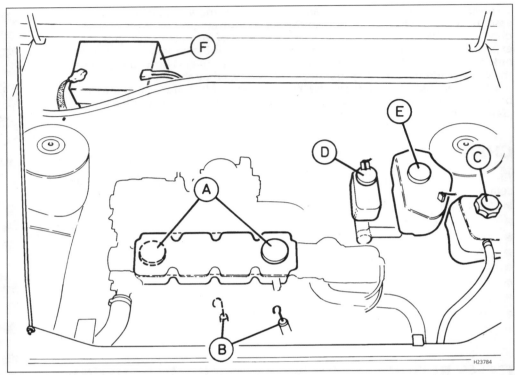

▲ *Typical engine compartment*
- **A** *Engine oil filler cap (alternative location also shown)*
- **B** *Engine oil dipstick (alternative location also shown)*
- **C** *Coolant expansion tank with filler/pressure cap*
- **D** *Brake fluid reservoir*
- **E** *Washer fluid reservoir*
- **F** *Battery*

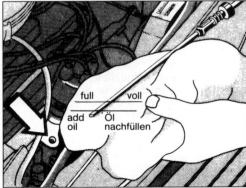

▲ *Checking the engine oil level on the 1.0 litre engine*

Checking oil level

To check the oil level, the car should be standing on level ground, and the engine should have been stopped for at least a few minutes. Open the bonnet and look for the dipstick, which is located on the front of the engine. On the 1.0 litre engine, it is located next to the distributor, but on all other engines it is located in a tube.

Pull out the dipstick, wipe it clean (with a clean non-fluffy cloth), then slowly push it fully back into its location and pull it out again. Note the oil level, which should never be allowed to drop below the bottom mark on the dipstick. If the oil needs topping-up, unscrew

VAUXHALL NOVA

REGULAR CHECKS 75

and remove the oil filler cap and top-up to the upper level mark on the dipstick. On models up to November 1990, the dipstick only has marks on it, but on later models the lower mark is indicated by the word 'MIN' and the upper mark by the word 'MAX'.

Note that the amount of oil required to raise the level on the dipstick from the lower to upper marks is 0.75 litre (1.3 pint) on 1.0, 1.2, 1.3 & 1.4 litre engines, and 1.0 litre (1.8 pint) on the 1.6 litre engine. Always try to use the same type and make of oil, and take care not to overfill. Mop up any oil which might have been spilt, and make sure that the oil filler cap is correctly refitted.

If the oil needs to be topped-up regularly, check for leaks, and if necessary seek advice.

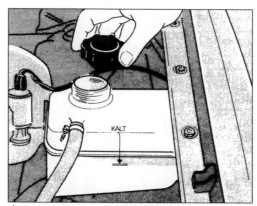
▲ *Removing the cap from the expansion tank – note the KALT (cold) level mark and the lower minimum mark*

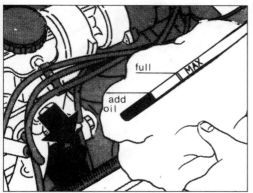
▲ *Checking the engine oil level on the 1.2 litre engine*

▲ *Topping-up the engine oil (1.0 litre engine model shown)*

Checking coolant level

The coolant level can be checked visually by checking the level in the expansion tank when the engine is **cold**. The level should be a little above the KALT (cold) mark.

If topping-up is necessary, first the cooling system pressure cap must be removed. Topping-up should always be done with the engine cold, but if it does prove necessary to remove the cooling system pressure cap with the engine hot for any reason, take care to avoid scalding. Place a thick rag over the cap, and loosen the cap slowly in stages to gradually release the pressure in the system.

A screw type pressure cap is fitted to the expansion tank. Slowly unscrew the cap until all the pressure in the system is released, then remove the cap.

Topping-up should always be done with a mixture of water and antifreeze to the same strength as the mixture already in the system (in this case, 55% clean water and 45% antifreeze). It's important to note that because of the different types of metals used in the engine, it's vital to use antifreeze with suitable anti-corrosion additives all year round.

Although plain water can be used for topping-up, it's unwise to make it a habit, as the strength of the antifreeze in the main system will gradually be diluted. **Never** use water alone to fill the whole system. After topping-up, refit and tighten the filler cap.

Normally, topping-up will rarely be required,

76 REGULAR CHECKS

▲ Topping-up the coolant through the expansion tank

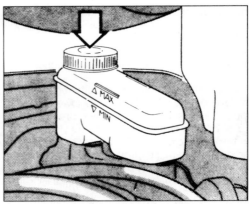

▲ Brake fluid reservoir level markings and filler cap (arrowed) on models up to November 1990

and if the need for regular topping-up arises, it will probably be due to a leak somewhere in the system. Leaks are most likely to occur from the radiator or the various hoses. If no leaks can be found, it's possible that there's an internal fault in the engine, such as a crack in the cylinder head, or a blown cylinder head gasket, but in this case it's best to seek specialist advice.

Checking brake fluid level

Note: *Refer to 'Safety first' on page 86 for the special precautions which should be taken when handling brake fluid.*

Although a low brake fluid level warning light is fitted to some models, the fluid level should always be checked visually whenever the oil and coolant levels are checked. The level should be up to the 'MAX' mark on the side of the reservoir, but it is quite normal for the level to fall slightly as the brake friction material wears.

The fluid level must **never** be allowed to drop below the 'MIN' mark.

If topping-up is required, always use the correct type of fluid, which should always be stored in a full, airtight container. Don't top-up using fluid which has been stored in a partly-full container, as it will have absorbed moisture from the air, which can dangerously reduce its performance.

Topping-up should hardly ever be required, unless there's a leak somewhere in the hydraulic system. If a leak is suspected, the car

▲ Brake fluid reservoir level markings and filler cap on models from November 1990 onwards

should not be driven until the braking system has been thoroughly checked. **Never** take any risks where brakes are concerned.

Checking tyres

It's extremely important to carry out regular checks on the tyres, to make sure that the pressures are correct, and that the tyres are not damaged. The tyres are the only part of the car in contact with the road, so their condition will affect the steering and general handling of the car, and therefore its safety.

To check tyre pressures accurately, the tyres must be cold, which means that the car must not have been driven recently. Note that it can make a noticeable difference to the tyre

VAUXHALL NOVA

REGULAR CHECKS 77

pressures if the car has been standing out in the sun. This can be very noticeable when one side of the car is in the sun, and the other side is in shadow.

Note that the recommended tyre pressures vary depending on the size of tyres fitted – refer to *'Service specifications'* on page 71. The size of the tyre (eg 155 SR 13) is clearly marked on the tyre sidewall, although the style of the tyre size marking may vary depending on the tyre manufacturer – consult a Vauxhall dealer or a tyre specialist for further details of tyre size markings, and the latest pressure recommendations.

If the tyres are being checked after the car has been driven, for example when filling up with petrol during a journey, the pressures are bound to be higher than specified. It's best to simply check that the pressures in the two front tyres are *equal*, and similarly for the two back tyres (remember that the specified pressure for the rear tyres may be different to the front).

Never let air out of the tyres of a car which has recently been driven, to bring the pressures down to that specified (unless the tyre pressures had been increased for a fully-loaded car, and the load has just been reduced).

When checking the tyre pressures, don't forget to check the spare!

With the tyres properly inflated, run your fingers around the edge of the tyre to check for any cuts and bulges in the tyre walls. Ideally, the car should be jacked up to visually check the tyres all round but, alternatively, the car can be rolled backwards and forwards to enable you to check all round the tread. Check the tread for cuts, and remove any debris, such as small stones and broken glass, using a screwdriver or similar tool. Any large items such as nails which have penetrated the surface of the tyre should be left in the tread (to identify the location of the damage) until

Shoulder wear
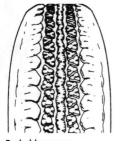

Probable cause:
Underinflation (wear on both sides)
Action: Check and adjust pressure

Probable cause:
Incorrect wheel camber (wear on one side)
Action: Repair or renew suspension parts

Probable cause:
Hard cornering
Action: Reduce speed

Centre wear
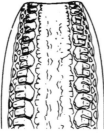

Probable cause:
Overinflation
Action: Measure and adjust pressure

Uneven wear
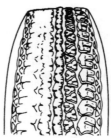

Probable cause:
Incorrect camber or castor
Action: Repair or renew suspension parts

Probable cause:
Malfunctioning suspension
Action: Repair or renew suspension parts

Probable cause
Unbalanced wheel
Action: Balance tyres

Probable cause:
Out-of-round brake disc/drum
Action: Machine or renew disc/drum

Toe wear
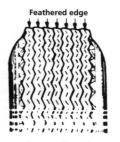

Probable cause:
Incorrect toe setting
Action: Adjust front wheel alignment

▲ *Tyre wear patterns and causes*

VAUXHALL NOVA

REGULAR CHECKS

the tyre has been repaired. Refer to *'Breakdowns'* on page 44 for details of how to fit the spare wheel.

Also check the tyre treads for wear. Uneven wear across one particular tyre may be due to a fault with the suspension or steering. By law, the tread depth must be at least 1.6 mm (0.06 in), throughout a continuous band comprising the central three-quarters of the width of the tyre tread, around the full circumference of the tyre.

If you find that any of the tyres is excessively worn or damaged, obtain a new tyre as soon as possible. It's a good idea to try and stick to one type of tyre if possible, rather than mixing several different makes on the car. Generally, you'll find that it's much cheaper to buy tyres from a tyre specialist, rather than an ordinary garage. Make sure that when you buy a new tyre you have the wheel balanced, otherwise you might find that the new tyre causes vibration when driving.

Checking washer fluid level

The fluid reservoir for the windscreen washer, and (where applicable) for the tailgate rear window washer, is located on the left-hand side of the engine compartment (when viewed from the driver's seat).

If a headlamp washer is fitted, the fluid reservoir is located in the right-hand front corner of the engine compartment (when viewed from the driver's seat).

If topping-up is necessary, pull the cap from

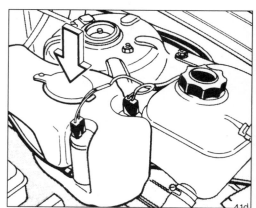
▲ *Windscreen washer/tailgate rear window washer fluid reservoir showing filler cap*

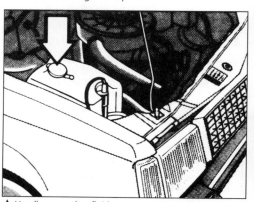
▲ *Headlamp washer fluid reservoir showing filler cap*

the reservoir, and top-up as necessary with clean water. A suitable washer fluid additive will keep the glass free from smears, and will prevent the fluid freezing in Winter.

Checking battery electrolyte level

The battery fitted as original equipment is of the maintenance-free type, which is 'sealed for life' and cannot be topped-up with distilled water. A maintenance-free type of battery is recognisable by not having any removable covers to enable topping-up (it will typically be marked 'Freedom Battery'). Where this type of battery is fitted, the electrolyte level and battery condition can be checked using the

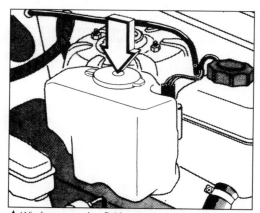
▲ *Windscreen washer fluid reservoir showing filler cap*

VAUXHALL NOVA

REGULAR CHECKS 79

battery condition indicator on top of the battery (this should be green if all is well). If the level is too low the battery will have to be replaced – refer to *'Breakdowns'* on page 47 for further information.

Where the original battery has been replaced by a standard type or by a low-maintenance type battery, the electrolyte level should be checked as follows.

The battery is located in the rear right-hand corner of the engine compartment (when viewed from the driver's seat) on the bulkhead. Remove the caps or the cover from the top of the battery, and check the level of the electrolyte fluid (with some types of battery, you can see the fluid level through the battery case). The level should be just above the tops of the plates inside the battery, or (where applicable) up to the mark on the battery case.

If necessary, add distilled water (**not** ordinary tap water) to each cell, to bring the level just above the tops of the battery plates, or up to the mark, as applicable.

With some types of battery, distilled water is added to a trough in the top of the battery until all the filling slots are full, and the bottom of the trough is just covered. On completion, refit the caps or cover, and carefully wipe up any drops of water that were spilt.

Topping-up should hardly ever be required, and the need for frequent topping-up indicates that the battery is being overcharged due to a fault in the charging circuit – seek advice from someone suitably qualified if this is the case.

Checking wipers and washers

Check the wipers and washers to make sure that they're working properly (don't forget the rear wiper and washer, on Hatchback models). Don't allow the wipers to work on dry glass for too long, as it will strain the motor and wear out the wiper blades. If necessary, the washer nozzles can be adjusted using a pin inserted into the end of the nozzle. If you're adjusting the windscreen washer nozzles, remember to aim them fairly high on the windscreen, as the airflow will usually deflect the spray down when the car is moving.

Over a period of time, the wiping action of the wiper blades will deteriorate, causing smearing of the glass. The rubber may also crack, particularly at the edges of the blades.

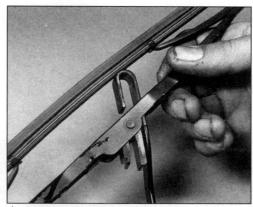

▲ *Removing the wiper blade from its arm*

When this happens, the blades must be renewed. This is a straightforward job, and is accomplished as follows:

Pull the wiper arm away from the glass, against the spring pressure, until the arm clicks into position. Swivel the blade on the arm, then depress the catch on the U-shaped retainer, and slide the blade down, then up from the wiper arm. Take care not to allow the arm to snap back against the glass.

The blade is normally renewed complete; the rubber insert can be renewed separately, but this is a very fiddly job.

Checking lights and horn

Switch on all the lights in turn, and check that they're working. Don't forget to check the direction indicators and the brake lights (the ignition must be switched on to check these), either with the help of an assistant, or by looking for the reflection in a suitable window or door.

Check that the direction indicators work with the brake lights on, and that the brake lights work with the tail lights on. Some faults may cause the various rear lights to interact, which can be dangerous, as it may confuse or distract following drivers.

Check the operation of the horn. A short 'blast' should be enough to prove that it at least works, but from time to time check its operation over several seconds.

If any of the bulbs need renewing, refer to *'Bulb, fuse and relay renewal'* on page 113. Any other problems are likely to be caused by a

REGULAR CHECKS

faulty switch, or loose or corroded connections (especially earth connections), and it's best to seek advice to solve these.

Checking for fluid leaks

Check the ground where the car is normally parked for any stains or spots of fluid which may have been caused by fluid leaking from the car.

Open the bonnet, and make a quick check of all the hoses and pipes, and the surfaces of all the components in the engine compartment. If there's any sign of leaking fluid, try to find the source of the leak, and seek advice if necessary. It can be difficult to identify leaking fluids, but if there's obviously a major leak, or if you suspect even a slight brake fluid or petrol leak, don't drive the car until the problem has been investigated by someone suitably qualified.

VAUXHALL NOVA

SERVICING

Regular servicing will ensure that your car is reliable and safe to drive, and could save you a lot of money in the long run. Many of the unexpected expenses which can crop up if things go wrong with your car can be avoided by carrying out regular servicing.

A significant proportion of the average garage servicing bill (well over 50% in many cases) is made up of labour costs, so obviously a lot of money can be saved by carrying out the work yourself. You'll find that most of the servicing jobs are very straightforward, and you don't need to be a mechanical genius to 'have a go' yourself. Don't be put off when you look under your bonnet; there are surprisingly few items which require frequent attention, and most of those which do are easily accessible without the need for anything more than basic tools. You'll discover that DIY servicing can be very rewarding, and will help you to understand what makes your car 'tick'.

The following chart lists all the servicing tasks recommended by the car manufacturer, and details of how to carry out the necessary work can be found in the subsequent pages. A few of the jobs require more extensive knowledge, or the use of special tools, and detailed explanation is beyond the scope of this Handbook. These more complicated tasks are identified on the chart, and details can be found in our Owners Workshop Manual for your car. Even if you decide to have the work done by a garage, the chart will enable you to check that the necessary work has been carried out.

Whenever you're carrying out servicing, safety must always be the first consideration, and you should read through the *'Safety first!'* notes on page 86 before proceeding any further.

SERVICE SCHEDULE

ENGINE COMPARTMENT
Check all components for fluid leaks, corrosion or deterioration

ENGINE
Check engine oil level

Renew engine oil and filter

On 1.0 litre engines, check valve clearances

Check/adjust timing belt (except 1.0 litre engine)

COOLING SYSTEM
Check coolant level

Check antifreeze concentration

FUEL AND EXHAUST SYSTEMS
Lubricate throttle/choke control linkages

Clean fuel pump filter (early models)

On 1.0 litre engines, clean the fuel inlet filter

Check exhaust system

Renew air cleaner filter element

Check idle speed/mixture settings

Check exhaust emissions

Check catalytic converter (where fitted)

On fuel injection engines, renew the fuel filter

IGNITION SYSTEM
On 1.0 litre engines, renew contact breaker points

Check/renew spark plugs

On 1.0 litre engines, check ignition timing

CLUTCH
Check clutch pedal adjustment

MANUAL GEARBOX
Check gearbox oil level

Note: *In addition to the above, also carry out the following task:*
Every 2 years, regardless of mileage, *renew the coolant (refer to page 104).*

VAUXHALL NOVA

SERVICE SCHEDULE

Regular Weekly	9000 miles (15 000 km) 12 months	18 000 miles (30 000 km) 2 years	36 000 miles (60 000 km) 4 years
■	■	■	■
■	■	■	■
	■	■	■
	☐	☐	☐
			☐
■	■	■	■
	■	■	■
	■	■	■
	■	■	■
	■	■	■
	■	■	■
		■	■
	☐	☐	☐
	☐	☐	☐
	☐	☐	☐
		☐	☐
	■	■	■
		■	■
	☐	☐	☐
		■	■
		■	■

Items marked ☐ are considered to be beyond the scope of this Handbook, and details can be found in the Owners Workshop Manual for your car (OWM 909).

VAUXHALL NOVA

SERVICE SCHEDULE

BRAKING SYSTEM
Check brake fluid level

Check front brake pads

Check/adjust rear brakes

Check handbrake and adjustment

SUSPENSION AND STEERING
Check tyres for condition and pressure

Check roadwheel bolts

Check rear wheel bearing adjustment

INTERIOR AND BODYWORK
Lubricate hinges/locks

ELECTRICAL SYSTEM
Check washer fluid level

Check battery electrolyte level (where applicable)

Check wipers and washers

Check operation of all lights and electrical equipment

Check alternator drivebelt

Check headlamp beam alignment – refer to a Vauxhall dealer for accurate setting

CAR UNDERSIDE
Check all components for fluid leaks, corrosion or deterioration

ROAD TEST
Check instruments and electrical equipment

Check steering, suspension and general handling

Check engine, clutch, gearbox and driveshafts

Check braking system

Note: *In addition to the above, also carry out the following task:*
Every 12 months, regardless of mileage, *renew the brake hydraulic fluid (refer to the* **Owners Workshop Manual**).

VAUXHALL NOVA

SERVICE SCHEDULE

85

Regular Weekly	9000 miles (15 000 km) 12 months	18 000 miles (30 000 km) 2 years	36 000 miles (60 000 km) 4 years
■	■	■	■
	☐	☐	☐
		☐	☐
		☐	☐
■	■	■	■
	■	■	■
		☐	☐
	■	■	■
■	■	■	■
■	■	■	■
■	■	■	■
■	■	■	■
	■	■	■
	☐	☐	☐
	■	■	■
	■	■	■
	■	■	■
	■	■	■
	■	■	■

Items marked ☐ are considered to be beyond the scope of this Handbook, and details can be found in the Owners Workshop Manual for your car (OWM 909).

SERVICE SCHEDULE 2

VAUXHALL NOVA

SERVICING

SAFETY FIRST!

Professional motor mechanics are trained in safe working procedures. No matter how enthusiastic you may be about getting on with the job you have planned, take time to read this Section. Don't risk an injury by failing to follow the simple safety rules explained here. The following is a guide to encourage you to be 'safety conscious' as you work on your car.

Essential points

ALWAYS use a safe system of supporting your car when working underneath it. The car's own jack (or a single hydraulic jack) is never sufficient. Once you've raised the car, you should use a safe means of holding it up. Axle stands, ramps or substantial wooden blocks are good; concrete blocks and bricks may crack or disintegrate, and should not be used. Make sure that you locate the supports where you know the car won't collapse, and where they can't slip.

ALWAYS loosen wheel nuts and other 'high torque' nuts or bolts (as applicable for the task to be carried out) before jacking up your car, or it may slip and fall.

ALWAYS make sure that your car is in **Neutral** or in **Park** (in the case of automatic transmission), and that the handbrake is securely on before trying to start the engine.

UNLESS the engine is cold, **ALWAYS** cover the cooling system's pressure (expansion tank) cap with several thicknesses of cloth before trying to remove it. Release the pressure slowly or the coolant may escape suddenly and scald you.

ALWAYS wait until the engine has cooled down before draining the engine (and/or transmission) oil, or the coolant. Oil or coolant that is very hot may scald you. If you can touch the engine sump/transmission/radiator (as applicable) without discomfort, the engine has probably cooled sufficiently to avoid scalding.

ALWAYS allow the engine to cool before working on it. Many parts of the engine become very hot in normal operation, and you may burn yourself badly.

ALWAYS keep brake fluid, antifreeze and other similar liquids away from your car's paintwork, as it may damage the finish (or may remove the paint altogether!).

ALWAYS keep toxic fluids, such as fuel, brake and transmission fluids and antifreeze off your skin, and never syphon them by mouth.

ALWAYS wear a mask when doing any dusty work, or where spraying is involved, especially when doing body repair and brake jobs.

ALWAYS clean up oil and grease spills – someone may slip and be injured.

ALWAYS use the right tool for the job. Spanners and screwdrivers which don't fit properly are likely to slip and cause injury.

ALWAYS get help to lift and handle heavy parts – it's never worth risking an injury.

ALWAYS take time over the jobs you take on. Plan out what you must do, and make sure that you have the right tools and spare parts. Follow the recommended steps, and check over the job once you're finished. Leave 'short-cuts' to the experts.

ALWAYS wear eye protection when using power tools such as drills, sanders and grinders, when using a hammer, and when working beneath your car.

ALWAYS use a barrier cream on your hands when doing dirty jobs, especially when in contact with fuel, oils, greases, and brake and transmission fluids. It will help protect your skin against infection, and will make the dirt easier to remove later. Make sure that your hands aren't slippery. The use of a suitable specialist hand cleaner will make dirt and grease easier to remove without causing infection or damage to skin. Long term or regular contact with used oils and fuel can be a health hazard.

ALWAYS keep loose clothing and long hair out of the way of any moving parts.

ALWAYS take off rings, watches, bracelets

VAUXHALL NOVA

SERVICING 87

and neck chains before starting work on your car, especially the electrical system.

ALWAYS make sure that any jacking equipment or lifting tackle you use has a safe working load which will cope with what you intend to do, and use the equipment exactly as recommended by the manufacturers.

ALWAYS keep your work area tidy; someone may trip or slip on articles left lying around and be injured.

ALWAYS get someone to check up on you from time to time if you're working on the car alone, to make sure that you're alright.

ALWAYS do the work in a logical order, check that you've put things back together properly, and that everything is tightened as it should be.

ALWAYS keep children and animals away from an unattended car, and from the area where you're working.

ALWAYS park cars with catalytic converters away from materials which may burn, such as dry grass, oily rags, etc, if the engine has recently been run. Catalytic converters reach extremely high temperatures, and any such materials close by may catch fire.

REMEMBER that your safety, and that of your car and other people, rests with you. If you are in any doubt about anything, get specialist advice and help right away.

IF in spite of these precautions you injure yourself – get medical help immediately.

Asbestos

Some parts of your car, such as brake pads and linings, clutch linings and gaskets contain asbestos, and you should take appropriate precautions to avoid inhaling dust (which is hazardous to health) when working with them If in doubt, assume that they **do** contain asbestos.

Fire

Remember at all times that petrol is highly flammable. Never smoke, or have any kind of naked flame around, when working on the car. But the risk doesn't end there – a spark caused by an electrical short-circuit, by two metal surfaces contacting each other, by careless use of tools, or even by static electricity built up in your body under certain conditions, can ignite petrol vapour, which in a confined space is highly explosive.

If a fuel leak is suspected, try to find the cause; seek advice as soon as possible, and never risk fuel leaking onto a hot engine or exhaust. Catalytic converters (where fitted) run at extremely high temperatures, and therefore can be an additional fire hazard – observe the precautions outlined previously in this Section.

It's recommended that a fire extinguisher of a type suitable for fuel and electrical fires is kept handy in the garage or workplace at all times.

Never try to extinguish a fuel or electrical fire with water.

Fumes

Certain fumes are highly toxic, and can quickly cause unconsciousness and even death if inhaled to any extent, especially if inhalation takes place through a lighted cigarette or pipe. Petrol vapour comes into this category, as do the vapours from certain solvents such as trichloroethylene. Any draining or pouring of such fluids should be done in a well-ventilated area.

When using cleaning fluids and solvents, read the instructions carefully. Never use materials from unmarked containers – they may give off poisonous vapours.

Never run a car engine in an enclosed space such as a garage. Exhaust fumes contain carbon monoxide which is extremely poisonous; if you need to run the engine, always do so in the open air, or at least have the rear of the car outside the workplace. Although cars fitted with catalytic converters produce far less toxic exhaust gases, the above precautions should still be observed.

If you're fortunate enough to have the use of an inspection pit, never run the engine while the car is standing over it; the fumes, being heavier than air, will concentrate in the pit with possibly lethal results.

The battery

Never short across the two poles of the battery! (A battery is shorted by connecting the two terminals directly to each other, and this can happen accidentally when working under the bonnet with metal tools.) The heavy discharge caused will create 'gassing' of hydrogen from the battery, and this is highly explosive. Shorting across a battery may also cause sparks, and the combination can cause

VAUXHALL NOVA

SERVICING

the battery to explode, with potentially lethal results. A conventional battery will normally be giving off a certain amount of hydrogen all the time, so for the same reasons given above, never cause a spark or allow a naked flame close to it.

Batteries which are sealed for life require special precautions which are normally outlined on a label attached to the battery. Such precautions usually relate to battery charging and jump starting from another vehicle (for details of jump starting refer to 'Breakdowns' on page 48).

If possible, loosen the filler plugs or covers when charging the battery from an external source. Don't charge at an excessive rate, or the battery may burst. Special care should be taken with the use of high charge-rate boost chargers to prevent the battery from overheating.

Take care when topping-up and when carrying the battery. The battery contains dilute sulphuric acid which is very corrosive. It will burn and may cause long-term damage if in contact with skin, eyes or clothes. Similarly, corrosive deposits around the battery terminals may be harmful.

Always wear eye protection when cleaning the battery, to prevent the corrosive deposits from entering your eyes.

Mains electricity and electrical equipment

When using any electrical equipment which works from the mains, such as an electric drill or inspection light, always ensure that the appliance is correctly connected to its plug and that, where necessary, it's properly earthed. Always use an RCD (Residual Current Device, a safety device incorporating a circuit breaker) when using mains electrical equipment. Don't use such appliances in damp conditions, and take care not to create a spark or apply excessive heat close to fuel or fuel vapour. Also make sure that the appliances meet the relevant national safety standards.

Ignition HT voltage

A severe electric shock can result from touching certain parts of the ignition system, such as the spark plug HT leads, when the engine is running or being cranked, particularly if components are damp or the insulation is faulty. Where an electronic ignition system is fitted, the HT voltage is much higher than that used in a contact breaker system, and could prove fatal, especially to wearers of cardiac pacemakers.

Disposing of used engine oil

Used engine oil is a hazard to health and the environment, and should be disposed of safely and cleanly. Most local authorities provide a disposal site which will have a special tank for waste oil. If in doubt, contact your local council for advice on where you can dispose of oil safely – there may be a local garage who will allow you to use their specialised oil disposal tank free of charge. Remember that it's not just the oil which causes a problem, but empty containers, old oil filters, and oily rags, so these should be taken to your local disposal site too.

Do not under any circumstances pour engine oil down a drain, or bury it in the ground.

Jacking and vehicle support

The jack provided with the car is designed for emergency wheel changing, and should not be used for servicing and overhaul work. Instead, a more substantial workshop jack (trolley jack or similar) should be used. Whichever type is used, it's essential that additional safety support is provided by means of axle stands designed for this purpose. Never use makeshift means such as narrow wooden blocks or piles of house bricks, as these can easily topple or, in the case of bricks, disintegrate under the weight of the car. Further information on the correct positioning of the jack and axle stands is provided at the end of this Section.

If you don't need to remove the wheels, the use of drive-on ramps is recommended. Ensure that the ramps are correctly aligned with the wheels, and that the car is not driven too far along them, so that it promptly falls off the other end, or tips the ramps.

JACK AND AXLE STAND POSITIONS

When using a trolley jack or any other type of workshop jack to raise the car, the jack head should be placed under the special platforms located between the car jacking points and the wheel arch. To avoid damaging the underbody, it is recommended that a rubber pad or

SERVICING 89

▲ Trolley jack position for raising the front of the car

▲ Trolley jack position for raising the rear of the car

BUYING SPARE PARTS

When buying spare parts such as an oil or air filter, make sure that the correct replacement parts are obtained for your particular car. Many changes are made during the production run of any car, and when ordering spare parts, it will usually be necessary to know the year the car was built, the model type (eg 'L', 'SR' or 'GTE') and engine size, and in some cases the Vehicle Identification Number (VIN). The VIN is stamped on a metal plate attached to the cross panel at the front of the engine compartment, and is also stamped in the floor beneath the carpet next to the driver's seat. It also appears on the Vehicle Registration Document.

All of the components required for servicing will be available from an official Vauxhall dealer, but most parts should also be available from good motor accessory shops and motor factors.

Identifying engines

There are two types of engine fitted to the Nova – the overhead valve (OHV) 1.0 litre, and the overhead camshaft (OHC) 1.2, 1.3, 1.4 & 1.6 litre. On the 1.0 litre engine, the distributor is located vertically on the front of the engine block with its cap facing upwards. On the overhead camshaft engines, the distributor is located horizontally on the end of the engine, with its cap facing the cooling system expansion tank. In addition, the 1.2, 1.3 & 1.4 litre engines carry a prominent 'OHC' marking on top of the engine.

Distinguishing between a fuel injection engine and a carburettor engine is easy – fuel injection is only fitted to 1.6 litre GTE (or GSi) models, and to 1.4 litre catalytic converter-equipped models. The 1.6 litre engine is clearly marked '1.6 INJECTION' on top of the engine. Those 1.4 litre models fitted with a catalytic converter will carry badging to that effect on the rear of the car, and will only accept unleaded petrol. Under the bonnet, 1.4 litre catalyst models will *not* have a fuel pump of the type shown in the underbonnet illustration for the 1.2 litre engine.

wooden block is positioned between the jack head and the body.

Do not jack the car under the engine sump, transmission or any of the steering or suspension components.

Axle stands can be positioned in the same places, provided that a wooden or rubber pad is used to prevent damage to the sill. If necessary, place the trolley jack slightly to one side when jacking up the car, in order to leave room for the axle stand head when the car is lowered. Refer to the Owners Workshop Manual for details of alternative axle stand positions. Do not under any circumstances position axle stands under the rear axle centre section.

VAUXHALL NOVA

90 SERVICING

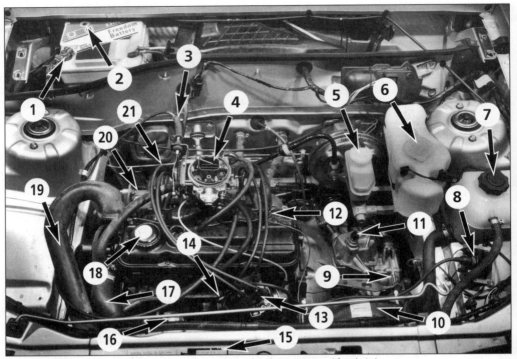

▲ Underbonnet component locations on a 1.0 litre engine (air cleaner removed for clarity)

1. Battery earth terminal
2. Battery condition indicator
3. Throttle cable
4. Carburettor
5. Brake fluid reservoir
6. Windscreen/tailgate window washer reservoir
7. Cooling system expansion tank filler/pressure cap
8. Ignition coil
9. Clutch cable adjustment nut
10. Radiator cooling fan shroud
11. Gearbox breather/oil filler plug
12. Heater hose connection for cooling system bleeding
13. Distributor cap and HT leads
14. Engine oil dipstick
15. Vehicle Identification Number (VIN) plate
16. Fuel pump
17. Radiator bottom hose
18. Engine oil filler cap
19. Radiator top hose
20. Alternator
21. Number 1 spark plug HT lead

VAUXHALL NOVA

SERVICING 91

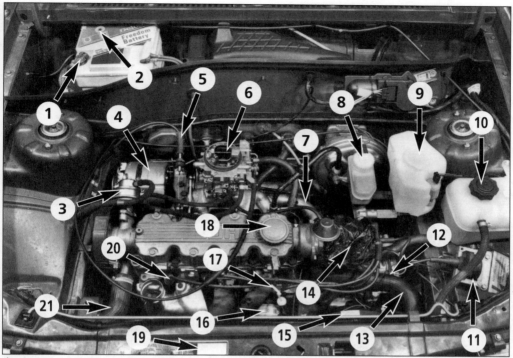

▲ Underbonnet component locations on a 1.2 litre engine (air cleaner removed for clarity) – 1.3 & 1.4 litre models similar

1 Battery earth terminal
2 Battery condition indicator
3 Fuel pump (not fitted to catalyst models)
4 Alternator
5 Throttle cable
6 Carburettor (throttle housing on catalyst models)
7 Inlet manifold
8 Brake fluid reservoir
9 Windscreen/tailgate washer reservoir
10 Cooling system expansion tank filler/pressure cap
11 Ignition coil
12 Clutch cable adjustment nut
13 Radiator bottom hose
14 Distributor cap and HT leads (under waterproof cover)
15 Radiator cooling fan shroud
16 Oil filter
17 Engine oil dipstick
18 Engine oil filler cap
19 Vehicle Identification Number (VIN) plate
20 Number 1 spark plug HT lead
21 Radiator top hose

VAUXHALL NOVA

92 SERVICING

▲ Underbonnet component locations on a 1.6 litre GTE model

1. Battery earth terminal
2. Battery condition indicator
3. Throttle housing
4. Throttle cable
5. Brake fluid reservoir
6. Windscreen/tailgate washer reservoir
7. Gearbox breather/oil filler plug
8. Cooling system expansion tank filler/pressure cap
9. Ignition coil
10. Radiator bottom hose
11. Distributor cap and HT leads
12. Radiator cooling fan shroud
13. Engine oil dipstick
14. Exhaust downpipes
15. Number 1 spark plug HT lead
16. Vehicle Identification Number (VIN) plate
17. Radiator top hose
18. Air cleaner
19. Airflow sensor
20. Engine oil filler cap

VAUXHALL NOVA

SERVICING

SERVICE TASKS

The following pages provide easy-to-follow instructions to enable you to carry out the regular service tasks recommended by the manufacturer.

Every 250 miles (400 km) or weekly

Refer to 'Regular checks' on page 73 and carry out all the checks described.

Every 9000 miles (15 000 km) or 12 months – whichever comes first

ENGINE COMPARTMENT

Check all hoses and pipes, and the surfaces of all components for signs of fluid leakage, corrosion or deterioration

If there's any sign of leaking fluid, try to find the source of the leak, and seek advice if necessary. It can be difficult to identify leaking fluids, but if there's obviously a major leak, or if you suspect even a slight brake fluid or petrol leak, don't drive the car until the problem has been investigated by someone suitably qualified.

Check all rubber or fabric hoses for signs of cracking, damage due to rubbing on other components, and general deterioration. It's sensible to have any suspect hoses renewed as a precaution against possible failure.

Check for obvious signs of corrosion on metal pipes, hose clips, and component joints, which may indicate a fluid leak. Any badly-corroded components should be cleaned and, if necessary, renewed. Pay particular attention to metal fluid pipes, and have them renewed if they're badly pitted or corroded.

ENGINE

Renew the engine oil and filter

Note: The following items will be required for this task:
- A suitable quantity of engine oil of the correct type – refer to **'Service specifications'** on page 69
- New oil filter (of the correct type for your particular car)
- New oil drain plug sealing washer (depending on the condition of the old one)
- Suitable container to catch the old oil as it drains (make sure that the capacity of the container is larger than the oil capacity of the engine – refer to **'Service specifications'** on page 70)
- Oil filter wrench – for removal of old oil filter
- Small quantity of clean rag
- Suitable spanner or socket to fit oil drain plug

The oil should be drained when the engine is warm, immediately after a run. Ideally, the front of the car should be jacked up to improve access. If the front of the car is to be raised, apply the handbrake, then jack up the front of the car and support it securely on axle stands (refer to 'Jacking and vehicle support' on page 88 for details of where to position the jack and axle stands).

Position a suitable container under the engine sump to catch all the oil which will be drained.

Remove the oil filler cap from the top of the engine, then working under the car, unscrew the drain plug from the engine sump using a suitable spanner or a socket. The drain plug is located on the bottom rear edge of the sump.

▲ Engine oil drain plug

SERVICING

Some oil will be released before the drain plug is removed completely, so take precautions against scalding, as the oil will be hot.

Try not to drop the drain plug as it is removed. Allow the oil to drain for at least 15 minutes.

Check the condition of the oil drain plug sealing washer, and renew it if necessary.

When the oil has finished draining, clean the drain plug, washer, and the mating face of the sump, then refit and tighten the drain plug. There's no need to strain yourself doing this, just make sure that the plug is tight. Remember that it could be you who has to undo it, next time!

Position a suitable container under the oil filter. On the 1.0 litre engine, the filter is located vertically on the front of the engine, below the distributor. On all other engines, the oil filter is located horizontally on the front of the engine, next to the exhaust downpipes – be careful not to burn yourself on the hot exhaust when removing the filter.

▲ *Oil filter location on engines except the 1.0 litre – viewed from under the car*

Unscrew the filter using a suitable chain or strap wrench to loosen it. As soon as it is removed, hold the filter upright, but be prepared for some spillage of warm oil. Make sure that the rubber sealing ring is removed with the old filter, and is not left on the engine.

Wipe clean the oil filter mounting flange on the engine. Smear a little clean engine oil on the sealing ring of the new oil filter, then screw the filter onto the engine, and tighten it *by hand only*. If no instructions are provided with the filter, tighten it until the sealing ring touches the mounting face on the engine, then tighten it a further three-quarters of a turn.

Where applicable, lower the car to the ground and, with the car parked on level ground, fill the engine with the correct quantity and grade of oil through the filler on top of the engine. Fill the engine slowly over a period of several minutes, until the level reaches the **'MAX'** or upper mark on the dipstick.

Make sure that the oil filler cap is fitted on completion.

Don't race the engine when starting it for the first time after an oil change, as the oil will take a few seconds to circulate. There may be a delay before the oil pressure warning light goes out, as the engine lubrication system fills with oil.

Run the engine, and check for leaks from the filter and the drain plug, then stop the engine and check the oil level. Refer to *'Regular checks'* on page 74 for details of how to check the oil level. Since some oil will have filled the oil filter, it will usually be necessary to top-up the level a little.

Dispose of the old engine oil safely (refer to *'Safety first!'* on page 88). **Don't** pour it down a drain.

COOLING SYSTEM

Check the coolant antifreeze concentration

Using an antifreeze hydrometer, withdraw some coolant and read off the concentration on the hydrometer float.

Refer to *'Regular checks'* on page 73 for the correct concentration, and if necessary adjust

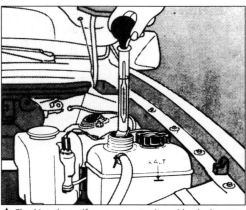

▲ *Checking the antifreeze concentration with a hydrometer*

VAUXHALL NOVA

SERVICING 95

by adding antifreeze. If this will raise the coolant level above the maximum mark, first withdraw some of the coolant using the hydrometer. If a large amount must be removed, drain off some of the coolant as described in the *'Every 2 years, regardless of mileage'* service on page 105.

FUEL AND EXHAUST SYSTEMS

Lubricate throttle/choke control linkages

Have an assistant operate the accelerator (and then the choke control, where applicable) while you look under the bonnet to identify the linkages. Apply a little engine oil to the throttle linkage on the carburettor or throttle housing, as applicable. Where applicable, lubricate the choke linkage on the carburettor. Operate the accelerator (and the choke control, where applicable) immediately afterwards to allow the oil to penetrate the linkages.

Clean the fuel pump filter screen (early carburettor models)

On 1.0 litre engines, the fuel pump is located on the front-facing side of the cylinder block at the right-hand end (when viewed from the driver's seat). On all other engines, the fuel pump is located on the rear of the engine, on the right-hand side (as viewed from the driver's seat). First unscrew and remove the screw from the centre of the fuel pump cover.

Withdraw the cover, then turn it over and remove the rubber sealing ring and filter screen.

Wipe out the cover and pump recess, and wash the filter screen in clean fuel if it is clogged or partially choked with dirt.

Refit the filter screen and rubber seal to the cover, then refit the cover to the pump body. Tighten the screw carefully – it is quite easy to strip the threads in the pump body.

Clean the fuel filter in the carburettor inlet (1.0 litre engine only)

Remove the air cleaner as described in the *'18 000 mile (30 000 km)'* service section on page 101 for removal of the spark plugs. Unscrew the fuel inlet union plug from the carburettor, and withdraw the gauze filter from the recess.

Clean the filter in fuel, then refit it and tighten the union plug. Refit the air cleaner.

▲ *Withdrawing fuel inlet filter*

Check the exhaust system

This check is best carried out when the car is raised and supported on axle stands, as it will then be much easier to view the complete exhaust system. Look for black sooty stains, especially at joints, which indicate that exhaust gases are escaping. Also look for severe rusting, and check the condition of the exhaust mountings. Where damage is evident, the exhaust system should be renewed or repaired, or the mountings renewed.

▲ *Removing the fuel pump cover, filter screen and rubber sealing ring*

VAUXHALL NOVA

SERVICING

IGNITION SYSTEM

Renew the contact breaker points (1.0 litre engine)

Note: *A set of feeler gauges will be required for this work.*

To get at the contact breaker points, you'll need to remove the distributor cap. Start by prising apart the waterproof cover, if fitted, and removing it. This exposes the distributor cap itself, which can then be removed by unscrewing the two screws or releasing the two spring clips, as applicable. Pull off the rotor arm from the distributor shaft.

On the Bosch type distributor, remove the plastic anti-condensation shield. On the Delco-Remy distributor, move the contact breaker arm spring blade away from the plastic insulator, and slip the low tension and condenser lead terminals off the insulator.

Unscrew the retaining screw securing the contact breaker plate to the distributor baseplate, and lift off the contact set.

▲ *Removing the rotor arm from the distributor shaft (Delco-Remy type)*

▲ *Removing the contact breaker plate retaining screw (Delco-Remy distributor)*

On the Bosch type distributor, disconnect the low tension lead from the spade terminal, then remove the screw and lift out the contact set from the distributor.

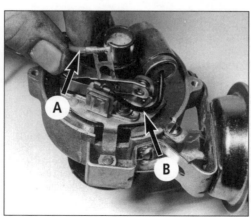

▲ *Slip the low tension (A) and condenser (B) leads out of the insulator (Delco-Remy distributor)*

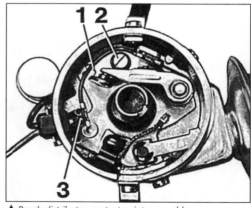

▲ *Bosch distributor contact points assembly*
  1 Notch for adjustment of gap
  2 Contact breaker set securing screw
  3 Low tension lead spade terminal

VAUXHALL NOVA

SERVICING 97

Clean the contact faces of the new points with methylated spirit before fitting.

Locate the new contact breaker set on the baseplate and refit the retaining screw, but do not fully tighten the screw at this stage.

On the Delco-Remy distributor, move the contact breaker spring blade away from the insulator, fit the low tension and condenser leads, and allow the spring blade to slip back into place. Make sure that the leads and the blade locate squarely in the insulator.

On the Bosch distributor, connect the low tension lead to the spade terminal.

The gap between the two contacts must now be adjusted, and for this, the engine must be turned. The engine may be turned by pushing the car along in top gear (manual gearbox), or by using a spanner on the crankshaft pulley bolt (the crankshaft pulley bolt is located on the lower right-hand end of the engine – viewed from the driver's seat – in the centre of the crankshaft pulley). Turning the engine will be much easier if the spark plugs are temporarily removed (refer to page xx), but be sure to mark the spark plug HT leads for position before removing them.

Turn the engine until the heel of the moving contact is resting on the peak of one of the cam lobes (the cam lobes are the four raised 'corners' on the distributor shaft).

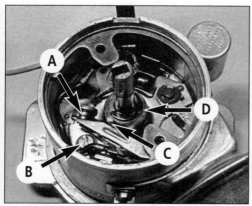

▲ Contact breaker points inside the Bosch distributor (Delco-Remy similar)
A Contact breaker points
B Securing screw
C Heel of moving contact
D Distributor shaft cam lobe

In this position, the gap between the contact point faces is at its maximum, and it is in this position that adjustment is made. A feeler blade equal to the contact breaker points gap as given in 'Service specifications' on page 70 should now just fit between the contact faces. The blade should be a firm sliding fit – neither too tight nor too loose.

▲ Checking the contact breaker points gap with a feeler blade (Delco-Remy distributor shown)

If the gap is not correct, move the breaker plate in or out to achieve the desired gap. The plate can be easily moved with a screwdriver inserted between the notch in the breaker plate and the two adjacent pips in the distributor baseplate. When the gap is correct, tighten the retaining screw and then recheck the gap. Lightly smear the surfaces of the cam with high-melting-point grease. Do not over-lubricate, as any excess could get onto the surfaces of the points and cause ignition malfunction. Apply one or two drops of engine oil to the felt pad in the top of the distributor shaft, then refit the rotor arm.

Before refitting the distributor cap, clean it thoroughly both inside and outside. Examine the inside of the distributor cap; if any of the four HT lead segments appear badly burnt or pitted, or if there are any signs of hairline cracks, renew the cap. Make sure that the carbon contact in the centre of the cap is not excessively worn, make sure that it's free to move, and that it protrudes from its holder – if not, renew the cap.

On the Bosch distributor, refit the plastic anti-condensation shield.

SERVICING

Refit the cap, and where necessary snap together the waterproof cover. Start the engine to make sure that the new points are correctly fitted.

On modern engines, setting the points gap as just described must be regarded as a preliminary adjustment. For best results, the dwell angle should now be measured. Since altering the points gap will also affect the ignition timing, this too should be checked after fitting a new set of points. For both these tasks, refer to the Owners Workshop Manual.

Check, and if necessary renew, the spark plugs

This work is not a Vauxhall requirement at this mileage and it is not therefore included in the service chart, but in order to maintain peak performance, it is highly recommended that it is carried out, particularly if the engine has completed a medium-to-high overall mileage. You will find the procedure described in the '18 000 mile (30 000 km)' service section on page 101.

SUSPENSION AND STEERING

Check the tightness of the roadwheel bolts

Using the brace from the toolkit, check the tightness of all roadwheel bolts.

BODYWORK

Lubricate all hinges, door locks, and the bonnet release mechanism

Use general-purpose light oil for the door locks, and a suitable general-purpose grease for the hinges and bonnet release mechanism.

Don't use an excess of lubricant, as it may find its way onto other components, or onto the clothing of the car's occupants!

ELECTRICAL SYSTEM

Check the condition and adjustment of the alternator drivebelt

Note: On the 1.0 litre engine, the drivebelt runs in the crankshaft, water pump and alternator pulleys. On all other engines, the drivebelt only runs in the crankshaft and alternator pulleys, as the water pump is driven by the toothed camshaft belt (timing belt). However, the removal, refitting and adjustment procedures on all models are similar.

Examine the drivebelt(s) for signs of cracking, obvious wear, or contamination, and renew if necessary.

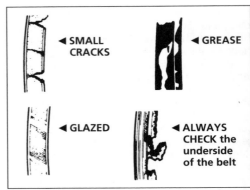

▲ Examine the alternator drivebelt for signs of wear, deterioration or contamination

To check the tension of the drivebelt, press on it midway between the alternator and crankshaft pulleys. The deflection of the drivebelt should be approximately 13.0 mm (0.5 in) under moderate finger or thumb pressure.

To adjust the alternator drivebelt tension, slacken the alternator mounting and adjuster link bolts just enough so that the alternator can be moved and will remain in its new position.

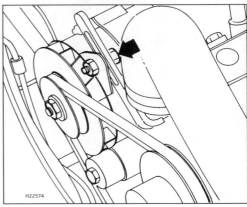

▲ Alternator upper adjuster link bolt (1.0 litre engine shown)

VAUXHALL NOVA

SERVICING

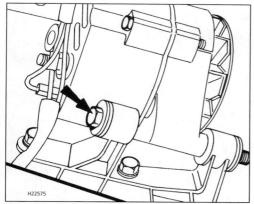

▲ *Alternator lower mounting (pivot) bolt*

Move the alternator towards or away from the engine as necessary to give the correct tension – if necessary, use a suitable lever. Once the tension is correct, tighten the mounting and adjuster link bolts.

To renew a drivebelt, slacken the mounting and adjuster bolts and move the alternator towards the engine, then remove the old drivebelt from the pulleys.

Fit the new drivebelt, and tension as already described. Run the engine for approximately ten minutes, then recheck the tension.

CAR UNDERSIDE

Check all hoses and pipes, and the surfaces of all components for signs of fluid leakage, corrosion, damage or deterioration

Refer to the *'ENGINE COMPARTMENT'* checks on page 93.

In addition, check the exhaust system and the fuel tank for any sign of damage or serious corrosion (light surface rust is to be expected).

Check the exhaust mountings to make sure that they're secure.

Check the visible suspension and steering components for obvious signs of damage or wear.

Check the driveshafts for obvious signs of distortion or damage.

Pay particular attention to the driveshaft and steering gear rubber gaiters, which should be renewed if they're split, or if there are any obvious signs of lubricant leakage. Also check the brake hydraulic hoses and pipes for damage or corrosion.

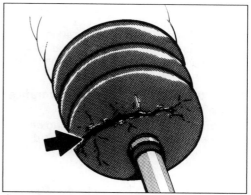

▲ *Check the driveshaft and steering gear rubber gaiters for splits (arrowed)*

ROAD TEST

Check the operation of all instruments and electrical equipment

Make sure that all instruments read correctly, and switch on all electrical equipment in turn to check that it functions properly.

Check for any abnormalities in the steering, suspension, handling or road feel

Drive the car, and check that there are no unusual vibrations or noises.

Check that the steering feels positive, with no excessive 'sloppiness' or roughness, and check for any suspension noises when cornering and driving over bumps.

Check the performance of the engine, clutch, gearbox and driveshafts

Listen for any unusual noises from the engine, clutch and gearbox.

Make sure that the engine runs smoothly when idling, and that there's no hesitation when accelerating.

Check that the clutch action is smooth and progressive, that the drive is taken up smoothly, and that the pedal travel is not excessive. Also listen for any noises when the clutch pedal is depressed.

SERVICING

Check that all gears can be engaged smoothly without noise, and that the gear lever action is not abnormally vague or 'notchy'.

Listen for a metallic clicking sound from the front of the car as the car is driven slowly in a circle with the steering on full lock. Carry out this check in both directions. If a clicking noise is heard, this indicates wear in a driveshaft joint, in which case seek advice, and renew the joint if necessary.

Check the operation and performance of the braking system

Make sure that the car does not pull to one side when braking, and that the wheels do not lock prematurely when braking hard.

Check that there's no vibration through the steering when braking.

Check that the handbrake operates correctly, without excessive movement of the lever, and that it holds the car stationary on a slope.

ADDITIONAL TASKS

Note that as well as the tasks described in the preceding paragraphs, the additional tasks given in the *'Service schedule'* chart on page 82 should also be carried out at this service interval. These tasks require more detailed explanation, or the use of special tools, and are considered beyond the scope of this Handbook. For details of these additional tasks, refer to the Owners Workshop Manual.

Every 18 000 miles (30 000 km) or 2 years – whichever comes first

In addition to the items in the *'9000 miles (15 000 km)'* service on page 93, carry out the following.

FUEL AND EXHAUST SYSTEMS

Renew the air cleaner filter element

Carburettor models

To remove the air cleaner element, first unscrew the three central screws.

Release the spring clips around the edge of the cover or, if spring clips are not fitted,

▲ *Removing the three central screws*

▲ *Releasing the spring clips from the air cleaner cover*

▲ *Removing the air cleaner filter element*

VAUXHALL NOVA

SERVICING 101

carefully prise around the lower edge of the cover with your fingers to release the retaining lugs.

With the cover removed, lift out the element and discard it.

Wipe inside the air cleaner body, being careful not to introduce dirt into the carburettor throat. Remember to clean the inside of the cover as well.

Fit the new element, then refit and secure the cover, observing any alignment marks or lugs.

Fuel injection models (1.4 litre catalyst)

The 1.4 litre catalyst engine is fitted with a single-point injection system, and has an air cleaner similar to that fitted to the carburettor engine. The air cleaner element renewal procedure is very similar.

Fuel injection models (1.6 litre)

On 1.6 litre fuel injection models, the air cleaner cover is part of the airflow sensor housing. To remove the air filter element, release the spring clips and lift off the air cleaner cover with airflow sensor attached. Take care when handling the airflow sensor, as it is fragile (and expensive!). Remove the air filter element and discard it.

Wipe clean the air cleaner body and cover, then fit the new element, making sure that its seal engages the cover recess. Refit and secure the cover by snapping the spring clips into place.

▲ Lifting off the air cleaner cover with airflow sensor attached

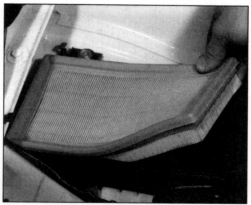

▲ Removing the air cleaner filter element

IGNITION SYSTEM

Check, and if necessary renew, the spark plugs

Note: *A suitable spark plug spanner, and a spark plug gap adjustment tool or a set of feeler gauges will be required for this task.*

It is preferable to remove the spark plugs when the engine is cold.

1.0 litre engines

On the 1.0 litre engine, the spark plugs are located on the rear of the engine, and it will be necessary to remove the air cleaner assembly complete for access to them. To do this, unscrew the three screws on the top of the air

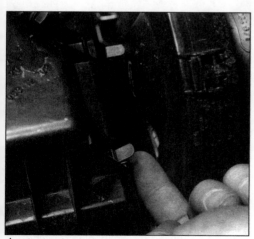

▲ Releasing the air cleaner cover spring clips

SERVICING

▲ Disconnecting the air cleaner spout from the hot air box supply tube on the exhaust manifold (1.0 litre engines only)

▲ Disconnecting a spark plug HT lead with metal shroud attached

cleaner cover, and lift the air cleaner from the top of the carburettor. At the same time, disconnect the air cleaner intake spout from the hot air box supply tube on the exhaust manifold (this just lifts off), and also disconnect the vacuum and breather hoses from the base of the air cleaner.

Remove the air cleaner assembly, and place it to one side.

All engines

If necessary, mark the HT leads using tape or sticky labels, to make sure that they're refitted in their correct positions.

If the plugs are removed one at a time, as described in the following procedure, there will be no chance of getting the leads wrong – work from one end of the engine to the other.

Starting with the first spark plug, look to see if a metal shroud is fitted at the end of the HT lead. These are normally fitted to three out of four of the plugs, on all engines except the 1.0 litre. If a shroud is fitted, remove the lead and shroud together by pulling on the metal shroud – this may be rather stiff, but careful pulling and twisting should free it. Do not use excessive force, or the spark plug will be damaged, and a new one would then be needed.

If there is no metal shroud, remove the HT lead by pulling on the rubber insulator at the end of the lead, not on the lead itself.

Before removing the spark plug, brush or blow any grit from the area around the plug recess to avoid it dropping into the engine as the plug is removed.

Using a spark plug spanner, unscrew the spark plug and remove it from the engine. The use of a spark plug spanner with a rubber insert will be helpful to remove the plug. Do not allow the tool to tilt, otherwise the ceramic insulator may be cracked or broken.

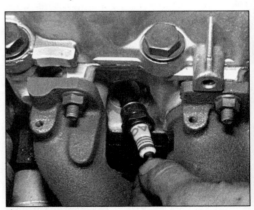

▲ Removing a spark plug

If the spark plug electrodes are heavily fouled (very dirty), it may be advisable to renew all four plugs, especially if poor starting has been experienced. A new set of spark plugs won't cost a great deal, but the improvement in engine performance is often dramatic, especially if several thousand miles were covered on the old set. Spark plug cleaning should only be carried out by someone suitably

SERVICING 103

qualified using specialised equipment, otherwise the plugs may easily be damaged.

If the plug appears to be in good condition, check the electrode gap as follows before refitting it (refer to *'Service specifications'* on page 70 for the correct electrode gap). The size of the spark plug electrode gap is extremely important, and it can drastically affect the performance of the engine.

Measure the gap between the electrodes using a spark plug gap adjustment tool, or a feeler gauge. The gap is correct when the tool or gauge is a firm sliding fit between the electrodes.

▲ *Checking the spark plug electrode gap*

▲ *Adjusting the spark plug electrode gaps*

If adjustment is required, carefully bend the **outer** electrode until the correct gap is obtained. **Never** try to bend the centre electrode.

Make sure that the plug threads and the seating area in the cylinder head are clean, and apply a little light oil to the plug threads. Check that the 'nut' on the threaded terminal on the top of the plug is tight, then screw the plug into the cylinder head by hand initially.

Using a spark plug spanner, tighten the plug as follows. With new plugs, screw in until the sealing washer contacts the cylinder head, then tighten further by no more than a quarter-turn. With plugs which have been used before, tighten by (approximately) a twelfth of a turn after the washer contacts the cylinder head, as the washer will already have been compressed. **Do not** overtighten the plugs.

Refit the HT lead (with metal shroud, where applicable), making sure that it is a secure fit over the end of the plug – there should be a soft 'click' as the lead is connected.

Repeat the above procedures for the remaining three spark plugs in turn, removing only one HT lead at a time, to avoid confusion.

On completion, check once more that all four HT leads are securely fitted. On the 1.0 litre engine, refit the air cleaner, first making sure that the rubber sealing ring is located correctly on the carburettor.

CLUTCH

Check the clutch pedal adjustment

With progressive wear of the clutch linings, the pedal will over a period of time move upwards towards the steering wheel, making adjustment necessary.

Unscrew the three retaining screws, and withdraw the parcel shelf from its location under the facia on the driver's side. Take a measurement from the edge of the steering wheel to the centre of the clutch pedal with the pedal in its normal released position (dimension **A**), and then take another measurement with the pedal fully depressed (dimension **B**). These measurements can be taken using a suitable strip of wood or metal, by making marks with the pedal released then depressed – the dimension you're after is the *difference* between the two measurements **A** and **B**.

The difference between the two measurements (B – A) is the pedal stroke, and this should be between 124.0 and 131.0 mm

VAUXHALL NOVA

SERVICING

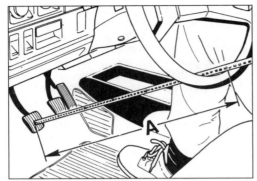

▲ *Clutch pedal released measurement (A)*

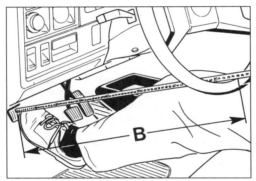

▲ *Clutch pedal depressed measurement (B)*

(4.9 and 5.2 in). If the pedal stroke is a long way off this, the nut on the threaded end of the cable, where it fits into the clutch release lever on the gearbox, must be adjusted. Remove the spring clip on the cable end fitting before making the adjustment.

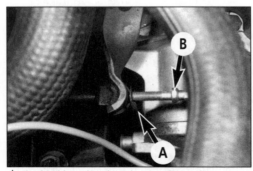

▲ *Clutch cable at the release lever on the gearbox*
A Spring clip   **B** Adjuster nut

Note that when correctly adjusted, the clutch pedal will be higher than the brake pedal – if the clutch pedal is aligned with the brake pedal, the cable is in need of adjustment. Also note that the design of the clutch means that there is no play at the pedal.

Recheck the pedal movement after making an adjustment, and do not forget to refit the spring clip.

MANUAL GEARBOX

Check and if necessary top-up the gearbox oil

The level plug is located just to the rear of the left-hand driveshaft on the differential housing.

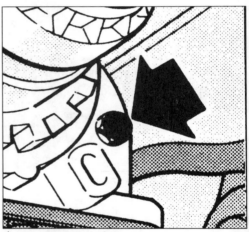

▲ *Manual gearbox oil level plug location*

Access to the level plug is only possible from under the car, so it is necessary to raise the car and support it on axle stands, or alternatively position the car over an inspection pit or on a ramp. Whichever method is used, it is important that the car is level.

Unscrew the level plug, and check that the oil level is up to the bottom of the plug hole by inserting a finger or a suitable length of wire.

If topping-up is required, unscrew and remove the breather plug from the top of the gearbox, and pour in the specified grade of oil (refer to *'Service specifications'* on page 69).

Do not under any circumstances top-up the gearbox through the level hole in the differential housing.

VAUXHALL NOVA

SERVICING 105

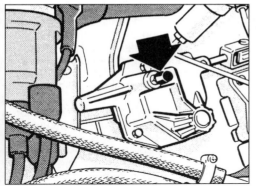

▲ Breather/oil filler plug location on the top of the gearbox

Note that if the gearbox requires repeated topping-up, there must be a leak somewhere, which should be attended to as soon as possible. Refit and tighten the filler/breather and level plugs on completion.

ADDITIONAL TASKS

Note that as well as the tasks described in the preceding paragraphs, the additional tasks given in the *'Service schedule'* chart on page 82 should also be carried out at this service interval. These tasks require more detailed explanation, or the use of special tools, and are considered beyond the scope of this Handbook. For details of these additional tasks, refer to the Owners Workshop Manual.

Every 2 years – regardless of mileage

COOLING SYSTEM
Renew the coolant

Note: *The following items will be required for this task:*
- *A suitable quantity of coolant solution (made up from approximately 55% clean water and 45% antifreeze)*
- *Suitable container to catch the old coolant as it drains (refer to* **'Service specifications'** *on page 70, and check that the container will be large enough for the volume of coolant in the system)*
- *Screwdriver to disconnect the radiator hoses*

Draining the system

The coolant should be drained with the engine cold.

Remove the cooling system pressure cap; refer to *'Regular checks'* on page 75 for details.

Position a suitable container beneath the radiator bottom hose connection, then slacken the hose clip, release the hose, and allow the coolant to drain. Position the bottom hose as low down as possible so that the maximum amount of coolant is drained. On 1.0 litre engines, use an Allen key to unscrew and remove the cylinder block drain plug (which is located on the rear-facing side of the engine, below the exhaust pipes) and allow the coolant to drain from the engine. Note that a certain amount of coolant will remain in the cylinder block.

If rust or sludge is evident in the coolant which has been drained, the cooling system should be flushed as described below, to remove any further contamination from the engine and radiator.

If the coolant drained is clean, then the cooling system can be refilled as described later in this Section.

Flushing the system

To flush the radiator, disconnect the top hose from the radiator, then insert a garden hose into the top of the radiator. Allow the water to circulate until it runs clear from the bottom of the radiator. It may be necessary to completely remove the radiator and invert it to be sure of cleaning away all sediment (refer to the Owners Workshop Manual).

The thermostat must be removed in order to flush the engine, but this work is outside of the scope of this Handbook. For those wishing to carry out the work, refer to the Owners Workshop Manual for the Nova (OWM 909).

Filling the system

Air must be released from the system as it is being filled. On the 1.0 litre engine, slacken the heater hose clip at the connection on the cylinder head and loosen the hose. On 1.2, 1.3 & 1.4 litre engines, disconnect the lead from the coolant temperature sender, and partially unscrew it from the inlet manifold. For better access, remove the air cleaner first (refer to page 100.

106 SERVICING

▲ *Coolant temperature sender on 1.2, 1.3 & 1.4 litre engines*

On 1.6 litre engines, the coolant temperature sender is located beneath the distributor – disconnect the lead and partially unscrew the sender.

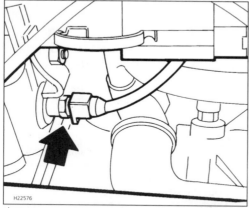

▲ *Coolant temperature sender on the 1.6 litre engine*

Make sure that all hoses are reconnected and their clips tightened; on 1.0 litre engines, refit and tighten the cylinder block drain plug.

A solution of 55% clean water and 45% antifreeze should be used to fill the system all year round. It's important to note that because of the different types of metals used in the engine, it's vital to use antifreeze with suitable anti-corrosion additives all year round. **Never** use water alone to fill the whole system.

Fill the system slowly through the expansion tank filler neck. Keep filling the system until coolant free of bubbles emerges at the temperature sender or heater hose (if necessary, temporarily disconnect the hose). Tighten the hose clip, or tighten the temperature sender and refit the lead, then refit the air cleaner (if removed).

Continue filling the system until the coolant is slightly above the KALT (cold) mark on the expansion tank. Squeezing the top and bottom hoses will help to dislodge air pockets in the system.

Screw on the expansion tank filler cap tight, then run the engine at a fast tickover, watching for leaks of coolant, until the electric cooling fan cuts in. Stop the engine and allow it to cool for at least two hours, then if necessary top-up the expansion tank so that the coolant is a little above the KALT (cold) mark. Remember that the cooling system must be cold in order to obtain an accurate level. Refit and tighten the filler cap.

Every 36 000 miles (60 000 km) or 4 years – whichever comes first

In addition to the items listed in the previous services, the additional engine task (for all except 1.0 litre engines) given in the *'Service schedule'* chart on page 82 should be carried out every 36 000 miles (60 000 km) or 4 years – whichever comes first. For details, refer to the Owners Workshop Manual.

SERVICING 107

SEASONAL SERVICING

If you carry out all the procedures described in the previous Section, at the recommended mileage or time intervals, then you'll have gone a long way towards getting the best out of your car in terms of both performance and long life. In spite of this, there are always other areas, not dealt with in regular servicing, where neglect can spell trouble.

A little extra time spent on your car at the beginning and end of every Winter will be well worthwhile in terms of peace of mind and prevention of trouble. The suggested tasks which follow have therefore been divided into Spring and Autumn Sections – but it's always a good idea to do them more frequently if you feel able.

Autumn

COOLING SYSTEM

Check the radiator and all hoses for signs of deterioration or damage

Refer to *'ENGINE COMPARTMENT'* in the *'9000 miles (15 000 km)'* service Section on page 93.

Check the strength of the coolant solution

Refer to *'COOLING SYSTEM'* in the *'9000 miles (15 000 km)'* service Section on page 94.

FUEL AND EXHAUST SYSTEM

Adjust the air cleaner setting (1.0 litre engine only)

The lever on the side of the air cleaner housing should be adjusted according to the outside air temperature as follows. Loosen the wing nut before making an adjustment, and tighten it afterwards. The temperatures are given on the air cleaner cover.

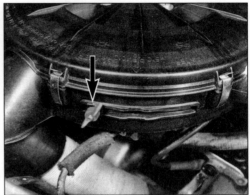

▲ *Air cleaner intake air temperature adjustment lever*
SUMMER   Above 10°C
INTERMEDIATE   10°C to -5°C
WINTER   Below -5°C

A check should be made of the setting throughout the year, to ensure that it corresponds with the appropriate outside temperature. No harm will come to the engine if the lever is incorrectly set for a while, but performance and fuel economy will suffer in the long run.

ELECTRICAL SYSTEM

Where applicable, check the battery electrolyte level

Refer to *'Regular checks'* on page 78.

Check and if necessary clean the battery terminals

Make sure that the terminals are secure, and clean any corrosion from the metal. Be sure to wear eye protection for this, as the white-coloured deposits are harmful. To prevent corrosion, the terminals can be coated with

VAUXHALL NOVA

petroleum jelly (don't use ordinary grease). It's worthwhile examining the battery tray at the same time, and if the same white-coloured deposits are present, clean them using a brush (or wash them away with plenty of hot water, being careful to avoid splashes).

Check the condition and adjustment of the alternator drivebelt

Refer to the procedure given in the *'9000 miles (15 000 km)'* service Section on page 98.

Check the operation of all lights, electrical equipment and accessories

Refer to *'Fault finding'* on page 128 if any problems are discovered.

If any of the bulbs need to be renewed, refer to *'Bulb, fuse and relay renewal'* on page 113.

Check the washer fluid level, and the wipers and washers

Refer to *'Regular checks'* on page 78.

TYRES

Check tread depth and condition

Refer to *'Regular checks'* on page 76. Remember that you may be driving in slippery conditions during the Winter.

BODYWORK

Thoroughly clean the car

Wash the car, and then polish it thoroughly to help protect the paint during the Winter.

Spring

UNDERBODY

Thoroughly clean the underside of the car

The best time to clean the underside of the car is after the car has been driven in wet conditions, when the accumulated dirt will be softened up.

To clean the car, first of all the car must be jacked up as high as possible (making sure that it is safely supported – refer to *'Safety first!'* on page 88).

Gather together a quantity of paraffin, or water-soluble solvent, a stiff-bristle brush, a scraper, and a garden hose.

With the car jacked up and safely supported, get underneath the car, and cover the brake components at each wheel with polythene bags to stop dirt and water getting into them.

Loosen any encrusted dirt, scraping or brushing it away – the paraffin or solvent can be used where there's oil contamination. Pay particular attention to the wheel arches. Take care not to remove the underseal (a waxy substance applied to the car underbody to protect against corrosion).

When all the dirt is loosened, a wash down with the hose will remove the remaining dirt and mud.

You can now check for signs of damage to the underseal. If there's any sign of the underseal breaking away, patch it up by spraying or brushing on a suitable underseal wax (available from motor accessory shops and motor factors). Make sure the area is clean and dry before applying the wax.

Take the opportunity to check for signs of rusting. Likely places are the body sills and floor panels.

If rust is found, seek advice, and have the affected area repaired before the problem gets too bad.

On completion, lower the car to the ground.

BODYWORK

Thoroughly check and clean the surfaces of the bodywork

Give the car a thorough wash, and check for stone chips and rust spots. For details of how to treat these minor paint problems, refer to *'Bodywork and interior care'* on page 111.

After any repairs to the paint have been carried out (not *before*, otherwise the paint will not stick), give the car a polish.

VAUXHALL NOVA

TOOLS

WHAT TO BUY

If you're intending to carry out your own servicing, you'll need to obtain a few basic tools. Although at first sight you may think that tools seem expensive, once you've bought them they should last a lifetime if you look after them properly.

The tools supplied with the car will enable you to change a wheel, and not much more. The absolute minimum tool kit you'll need to carry out any maintenance or servicing will be a range of metric spanners, two screwdrivers (one for crosshead or 'Phillips' type screws), and a pair of pliers. With a bit of ingenuity, these items should enable you to complete the more basic routine servicing jobs, but they won't allow you to do much else.

When buying tools, it's important to bear in mind the quality. You don't need to buy the most expensive tools available, but generally you get what you pay for. Cheap tools may prove to be a false economy, as they're unlikely to last as long as better quality alternatives.

It's very difficult to lay down hard and fast rules on exactly what you're going to need, but the following list should be helpful in building up a good tool kit. Combination spanners (ring one end, open-ended the other) are recommended because, although more expensive than double open-ended ones, they give the advantages of both types.

- Combination spanners to cover a reasonable range of sizes (say 7 mm to 17 mm at least)
- Adjustable spanner
- Spark plug spanner (with rubber insert)
- Spark plug gap adjustment tool
- Set of feeler gauges
- Screwdriver (Plain) – 100 mm long blade x 6 mm diameter (approx)
- Screwdriver (Crosshead) – 100 mm long blade x 6 mm diameter (approx)
- Oil filter wrench
- Pliers
- Tyre pump
- Tyre pressure gauge
- A suitable key (Torx or Allen type, depending on model) to fit the gearbox oil filler plug (later models only)
- Oil can
- Funnel (medium size)
- Stiff brush (for general cleaning jobs)
- Tool box
- Hydraulic jack (ideally a trolley jack)
- Pair of axle stands
- Suitable containers for draining engine oil and coolant
- Inspection lamp

This is by no means a comprehensive list of tools, and you'll probably want to gradually add to your tool box as you discover the need for other tools, especially if you decide to tackle some of the more advanced servicing jobs described in the **Owners Workshop Manual** for your particular car.

CARE OF YOUR TOOLS

Having bought a reasonable set of tools, it's worth taking the trouble to look after them. After use, always wipe off any dirt or grease using a clean, dry cloth before putting them away. Never leave tools lying around after they've been used – a tool rack, or better still a proper toolbox will prove the best way of keeping everything up together. Rags can be wrapped around loose tools to prevent them from rattling if you're going to keep them in the boot of the car.

Feeler gauges should be wiped with an oily cloth from time to time, to leave a thin coating of oil on the metal which will prevent corrosion. Screwdriver blades inevitably lose their keen edges, and a little occasional attention with a file or an oilstone will keep them in good condition.

VAUXHALL NOVA

110 BODYWORK AND INTERIOR CARE

VAUXHALL NOVA

BODYWORK AND INTERIOR CARE

It's well worth spending the time and effort necessary to look after your car's bodywork and interior. Cleaning your car regularly will not only improve its appearance, it will also protect the bodywork against the elements and the grime encountered in everyday driving. It's worth bearing in mind that if your car looks clean and tidy, it will also be worth more money when you come to sell it or trade it in for a newer model.

This Section will help you to keep your car in 'showroom' condition. If you wish to tackle repairs to more serious bodywork damage or corrosion, comprehensive details can be found in our **'Car Bodywork Repair Manual'**.

CLEANING THE INTERIOR

It's a good idea to clean the interior of the car first, before cleaning the exterior, as this will avoid spreading the dirt from inside over the bodywork.

Start by removing all the loose odds and ends from inside the car, not forgetting the ashtrays, glovebox and any interior pockets. Take out any loose mats or carpets, which should be shaken and brushed, and if possible vacuum-cleaned.

The inside of the car can be cleaned with a brush and dustpan, or preferably a vacuum-cleaner. If you can't use your household vacuum-cleaner, it may be worth investing in one of the small 12-volt hand vacuum-cleaners which can be powered from the car battery. If the carpets are very dirty, use a suitable proprietary cleaner with a brush to remove the ingrained dirt. Ideally, it's best to remove the carpets for cleaning, but this is an involved task in most modern cars, and it will probably be easier to leave them in place.

The facia, door trim, seats, and any other items which require attention can now be wiped over with a cloth soaked in warm water containing a little washing up liquid. If the trim is particularly dirty, use one of the proprietary cleaners available from motor accessory shops – a number of different types are available, suitable for vinyl, cloth or leather upholstery, etc, as required. An old nail brush or tooth brush will help to remove any ingrained dirt. On completion, wipe the surfaces dry using a lint-free cloth, and leave the windows open to speed up drying.

The inside of the windscreen and windows should be cleaned using a proprietary glass cleaner, or methylated spirits. Be careful about using certain household products such as washing-up liquid which may leave a smeary film. Finish off by wiping with a clean, dry paper tissue.

Don't forget to clean the boot.

Check for any nicks or tears in the interior trim, seats, headlining, etc. Various repair kits are available from motor accessory shops to suit most types of trim, but if the damage is very serious, the relevant trim panel will probably have to be renewed.

CLEANING THE BODYWORK

Ideally, the car should be washed every week, either by hand (preferably using a hosepipe), or by using a local car-wash. If you're washing the car by hand, use plenty of water to loosen the dirt and dust, and if possible use a suitable car shampoo or wax additive in the water. Any stubborn dirt such as road tar or bird droppings can be removed using methylated spirit or preferably one of the special proprietary cleaning solutions (in which case make sure that the product is suitable for use on car paint) – in either case, wash the affected area down with plenty of water after removing the dirt. **Never** just wipe over a dirty car, as this will scratch the paint very effectively.

VAUXHALL NOVA

BODYWORK AND INTERIOR CARE

Two or three times a year, a good silicone or wax polish can be used on the paintwork. It's important to wash the car and remove all stubborn dirt, tar, etc, before using polish, as the polish will effectively seal over the top of any dirt, making it extremely difficult to remove in the future. Good polishes actually form a protective coating over the paint finish, which should help to make future accumulated dirt easier to remove. Always follow the manufacturer's recommendations closely when using polish. Try to avoid getting polish on any unpainted plastic or rubber body parts such as bumpers and spoilers, as it tends to leave a stain when dry. Chrome parts are best cleaned with a special chrome cleaner, as ordinary metal polish will wear away the finish.

Plastic or rubber body parts such as bumpers and spoilers should be cleaned using a suitable proprietary cleaner. A number of different types are available, including colour restorers, and special solvents to remove any stray polish which may have crept onto the plastic when polishing the paintwork. Make absolutely sure that any solvents or cleaners used are suitable, as certain products may attack plastic and/or rubber.

If the paint is beginning to lose its gloss or colour, and ordinary polishing doesn't seem to solve the problem, it's worth considering the use of a polish with a mild 'cutting' action to remove what is in effect a surface layer of 'dead' paint. In this case, follow the manufacturer's instructions, and don't polish too vigorously, or you might remove more paint than you intended!

DEALING WITH SCRATCHES

With superficial scratches which don't penetrate down to the metal, repair can be very simple.

Very light scratches can be polished out by carefully using a suitable 'cutting' polish. Follow the manufacturer's instructions, and take care not to remove too much of the surrounding paint.

If the scratch cannot be polished out, touch-up paint will be required. Touch-up paint is usually available in the form of a touch-up stick from the car manufacturer's dealers, or in various forms from motor accessory shops. You will probably have to quote the year and model of the car in order to make sure that you obtain the correct matching colour, and in some cases you may have to quote a paint reference number which will usually appear on the car's Vehicle Identification Number (VIN) plate under the bonnet (refer to 'Buying spare parts' on page 89).

Lightly rub the area of the scratch with a very fine cutting paste to remove loose paint from the scratch and to clear the surrounding bodywork of polish.

Rinse the area with clean water.

Apply suitable touch-up paint to the scratch using a fine paint brush, and continue to apply fine layers of paint until the surface of the paint in the scratch is level with the surrounding paintwork.

Allow the new paint at least two weeks to harden, then blend it into the surrounding paintwork by rubbing the scratch area with a paintwork renovator or a very fine cutting paste.

If the paint finish requires a lacquer coat (in which case the lacquer will be supplied with the touch-up paint), it should now be applied to the newly painted area and allowed to dry in accordance with the manufacturer's instructions.

Finally, apply a suitable wax polish to the affected area for added protection.

VAUXHALL NOVA

BULB, FUSE AND RELAY RENEWAL

BULBS

A defective exterior light can be not only dangerous, but also illegal. Carrying spare bulbs will enable you to renew blown ones as they occur. A failed interior light bulb may be just a nuisance, but a faulty exterior light could be a life-or-death matter.

Before assuming that a bulb has failed, check for corrosion on the bulbholder and wiring connections, particularly around the rear light assemblies.

- Note that as a safety precaution, the battery earth (black) lead should always be disconnected before renewing a bulb.
- Remember to reconnect the lead on completion.

Bulb ratings

Always make sure that when a bulb is renewed, a new bulb of the correct rating is used.

All the bulbs are of the 12-volt type, and their ratings are as follows:

Bulb	Rating [watts]
Headlight	60/55
Side (parking) light	4
Side repeater light	5
Direction indicators	21
Stop/tail light	21/5
Number plate light	10
Reversing light	21
Interior and courtesy lights	10
Engine compartment light	10
Glove compartment light	10
Alternator (ignition) warning light	3
Instrument panel warning lights, ashtray light	1.2
Cigarette lighter illumination light	1.2
Instrument illumination	1.2 or 3
Switch illumination	0.5
Hazard flasher switch and heater control illumination	1.2
Front foglight	55
Rear foglight	21

BULBS, FUSES & RELAYS

VAUXHALL NOVA

114 BULB, FUSE AND RELAY RENEWAL

Exterior light bulbs

HEADLIGHT

Open the bonnet and engage the support. Disconnect the wiring plug from the rear of the headlight.

▲ *Disconnecting the headlight wiring plug*

▲ *Rubber cover removed to expose the spring clip*

Remove the rubber cover to expose the spring clip which secures the bulb.

Release the spring clips, then withdraw the bulb from the rear of the headlight and discard it; remember that it may be very hot if it has just been in use.

Fit the new bulb, being careful not to touch the glass of the bulb with the fingers. Clean the glass with methylated spirit or white spirit if the glass is accidentally touched, otherwise the grease from your fingertips may cause blackening and premature failure of the bulb. Make sure that the lugs on the bulb engage in the recesses in the headlight. Refit the rubber cover and wiring plug.

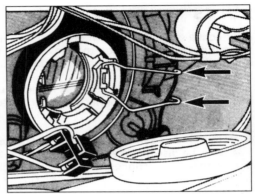

▲ *Rear of headlight with bulb removed, showing bulb retaining spring clips*

FRONT SIDELIGHT

Open the bonnet and engage the support. Reach down by the headlight and release the sidelight bulbholder from the headlight reflector by turning it anti-clockwise. Depress and twist the bulb to remove it from the bulbholder.

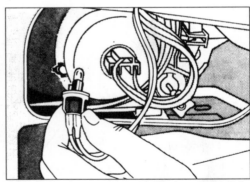

▲ *Removing a front sidelight bulbholder from the headlight*

Fit the new bulb, making sure that it is correctly engaged with the bulbholder, then insert the bulbholder into the headlight reflector and turn clockwise to lock.

FRONT DIRECTION INDICATOR LIGHT

Open the bonnet and engage the support.
On models up to November 1990, turn the bulbholder anti-clockwise and remove it from the rear of the direction indicator light. Fit the new bulb to the bulbholder, and refit the bulbholder to the light unit.

VAUXHALL NOVA

BULB, FUSE AND RELAY RENEWAL 115

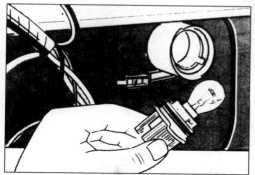

▲ Removing the front direction indicator light bulbholder on models manufactured up to November 1990

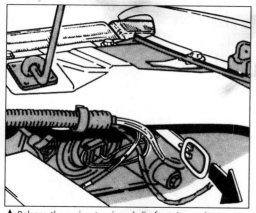

▲ Release the spring-tensioned clip from its retainer

▲ Withdrawing the front direction indicator light unit from its location pins

On models from November 1990 on, it is necessary to remove the direction indicator light unit before removing the bulbholder. To do this, pull the spring-tensioned clip and disengage it from the retainer, then withdraw the light unit directly forwards so that the location pins slide out of their sockets in the headlight unit.

Fit a new bulb to the bulbholder, refit the bulbholder, then refit the light unit, ensuring that the location pins and the spring-tensioned clip engage properly.

FRONT DIRECTION INDICATOR SIDE REPEATER LIGHT

Twist the light lens anti-clockwise and remove it, together with the rubber sealing ring.

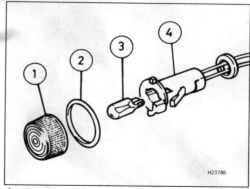

▲ Side repeater light
1 Lens   3 Bulb
2 Rubber ring   4 Bulbholder

Pull the bulb directly from its holder – it is of the wedge type, and must not therefore be turned. Push the new bulb into the holder, then refit the lens and rubber sealing ring.

FRONT FOGLIGHT

Unscrew the cross-head screw from the bottom of the front foglight, and lift the reflector and lens from the housing.

Release the two spring clips, and remove the bulb from the reflector.

Disconnect the feed wire at the connector. To prevent any damage to the lens and reflector, disconnect the earth wire also, then place the reflector in a safe position.

VAUXHALL NOVA

BULB, FUSE AND RELAY RENEWAL

▲ Removing the front foglight lens retaining screw

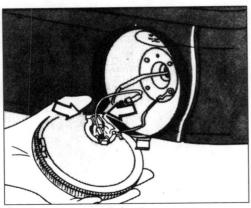
▲ Bulb retaining spring clip locations

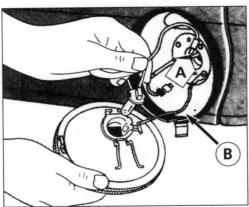
▲ Connector (A) on the feed wire leading to the bulb, and earth wire (B)

Fit the new bulb to the reflector, and engage the spring clips. Connect the new bulb wire and the earth wire, then refit the reflector and lens.

REAR LIGHT CLUSTER

Open the tailgate or bootlid, then unclip the trim panel from the rear corner of the luggage compartment. Press the plastic retaining lugs inwards, and remove the rear light cluster bulbholder assembly from the rear light unit.

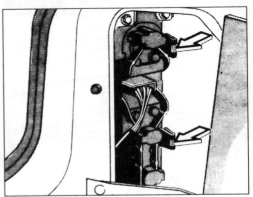
▲ Depress the plastic lugs on the rear light cluster bulbholder

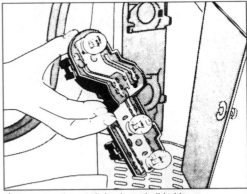
▲ Removing the rear light cluster bulbholder

Depress and twist the bulb to remove it from the bulbholder – the top bulb is the stop/tail light, the middle bulb is the direction indicator light, and the bottom bulb is the reversing light or foglight. Note that the stop/tail light bulb must be inserted in one position only – the engagement pins are offset to ensure correct

VAUXHALL NOVA

BULB, FUSE AND RELAY RENEWAL

fitting, and it is important not to force the bulb into its holder the wrong way round. Fit the new bulb, then refit the bulbholder and cover.

REAR NUMBER PLATE LIGHT

Press the number plate light to the left slightly, then use a small screwdriver to depress the clip

▲ *Insert a small screwdriver in the notch to release the number plate light assembly*

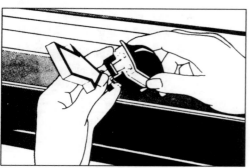
▲ *Separating the base from the lens*

▲ *Removing the number plate light bulb*

and prise the right-hand side of the light out of the bumper.

Separate the base of the light from the lens by depressing the protruding tongue.

Depress and twist the bulb to remove it from the light unit.

Interior light bulbs

Note: *Renewal of the various instrument panel and facia-mounted control illumination light bulbs requires detailed explanation, and in some cases extensive dismantling. Details can be found in the* **Owners Workshop Manual** *for your car (OWM 909).*

COURTESY LIGHT

Carefully prise the courtesy light from its location using a small screwdriver, taking care not to damage the surrounding trim, then pull the festoon type bulb from the spring contacts in the light body.

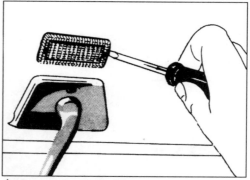
▲ *Prising out the courtesy light*

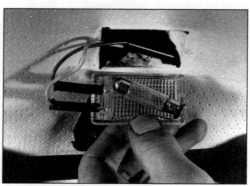
▲ *Festoon type bulb in the courtesy light*

VAUXHALL NOVA

118   BULB, FUSE AND RELAY RENEWAL

ENGINE COMPARTMENT/LUGGAGE COMPARTMENT LIGHT

Both the engine compartment and luggage compartment light bulbs are removed in the same way as the courtesy light bulb.

FUSES

Most of the car's electrical circuits are protected by fuses, which will blow when the relevant circuit becomes overloaded. This is to prevent possible damage to electrical components, or the possible risk of fire.

- Before renewing a fuse, always make sure that the ignition and the relevant circuit are switched off.

The fuses are located in a fusebox, mounted to the right of the steering column on the lower part of the facia. Access to the fuses is gained by removing the cover – pull on the bottom edge of the cover, then release the upper edge.

▲ Fuse locations on the inside of the cover

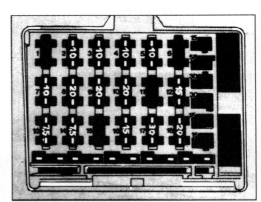
▲ Numbered fuse positions inside the fusebox

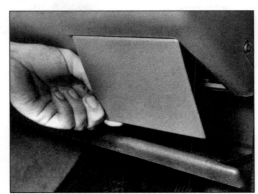
▲ Removing the fusebox cover

The fuse locations, current rating and circuits protected are shown on the inside of the cover, and the fuses are also numbered inside the fusebox.

All fuses are a push fit in their sockets, and a blown fuse can be recognised by a break in the wire connection between the two terminals.

If a fuse has blown, renew it with one of identical rating. To remove a fuse, simply pull it directly from the fusebox.

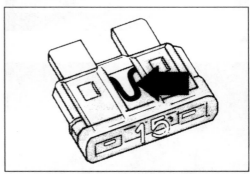
▲ A 15 amp fuse, showing the fuse wire which will melt when the circuit is overloaded

VAUXHALL NOVA

BULB, FUSE AND RELAY RENEWAL

▲ Removing a fuse from the fusebox

The fuses are colour-coded to show their rating as follows:

Colour	Rating (amps)
Brown	7.5
Red	10
Blue	15
Yellow	20
Green	30

Spare fuses may be inserted horizontally in the bottom row of the fusebox, and it is recommended that a complete selection is always carried in case of emergency. Always use proper fuses of the correct type and rating, and **never** use wire or any other material to bridge the gap where a fuse should be.

▲ Multi-tone horn fuse (below plastic cover) and relay

Note that the same fuse should never be renewed more than once without investigating the source of the trouble. Seek advice from someone suitably qualified if necessary.

Various additional in-line fuses may be used depending on any additional equipment fitted. On some later models fitted with a multi-tone horn, a fuse and relay is located on the left-hand side of the bulkhead, and the fuse is attached to the base of the relay.

If a problem is suspected which may be due to a fuse not located in the fusebox or bulkhead, it's best to seek advice from a Vauxhall dealer or auto electrician, who will know where to look.

RELAYS

Various types of relays are used in some circuits, and a faulty relay may prevent one or more components from working.

- **Before renewing a relay, always make sure that the ignition and the relevant circuit are switched off.**

There are three main relays, and they are located beneath the fusebox. They operate the windscreen wiper intermittent wipe delay (left relay), direction indicator/hazard flasher system (middle relay) and the heated rear window (right relay). Access is gained through the space between the lower edge of the facia and the parcel shelf. On models with a tailgate wiper, the intermittent wipe relay is located behind a quarter trim panel in the luggage compartment. On models with headlight washers, or an electric fuel pump (1.4 litre catalyst and 1.6 litre models only), the relays are located on the bulkhead in the engine compartment. On some later models fitted with a multi-tone horn, the relay is located on the left-hand side of the bulkhead. Additional relays may be fitted according to the vehicle equipment.

The relays are of the plug-in type, and are removed by simply pulling them out of their sockets.

It's difficult to tell visually whether a relay is faulty, and the best test is to substitute a known good relay to see if the relevant circuit then operates – if it does, then it's probably the relay which was causing the problem.

120 PREPARING FOR THE MOT TEST

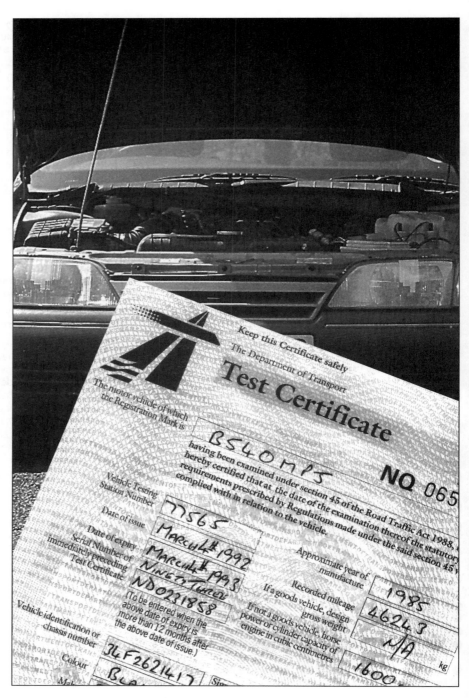

VAUXHALL NOVA

PREPARING FOR THE MOT TEST

The following information is intended as a guide to enable you to spot some of the more obvious faults which may cause your car to fail the MOT test.

Obviously it isn't possible to check the car to the same standard as a professional MOT tester, who will be highly experienced and will have all the necessary tools and equipment.

Although we can't cover all the points of the test, the following should provide you with a good indication as to the general condition of the car, and will enable you to identify any obvious problem areas before submitting your car for the test.

Further explanation of most of the checks, and details of how to cure any problems discovered can be found in the relevant *Owners Workshop Manual* for your particular car (OWM 909 Nova 1983 to February 1992).

Lights
- All lights must work reliably, and the lenses and reflectors must not be damaged.
- Pairs of similar lights must be of the same brightness (eg both rear lights).
- Both headlights must show the same colour, and must be correctly aimed so that they light up the road adequately, but don't dazzle other drivers.
- The brake lights must work when the brake pedal is pressed.
- The direction indicator lights must flash between one and two times per second.
- All switches and driver's tell-tale lights must be fitted, secure and in good working order.
- No lamp should be adversely affected by the operation of any other lamp.

Steering and suspension
- All the components and their mounting points must be secure, and there must be no damage or excessive corrosion.
- There must be no excessive free play in the joints and bushes, and all rubber gaiters must be secure and undamaged.
- The steering mechanism must operate smoothly without excessive free play or roughness.
- There should be no excessive free play or roughness in the wheel bearings, but there should be sufficient free play to prevent tightness or binding.
- There should be no major fluid leaks from the shock absorbers and, when each corner of the car is depressed, it should rise and then settle in its normal position.
- The driveshafts and propeller shaft (on rear wheel drive cars) must not be damaged or distorted.

Brakes
- It should be possible to operate the handbrake without excessive force or excessive movement of the lever, and it must not be possible for the lever to release unintentionally. The handbrake must be able to lock the relevant wheels on which it operates.
- The brake pedal must not be damaged, and when the pedal is pressed, resistance should be felt near the top of its travel – hard resistance should be felt, and the pedal should not move down towards the floor. The resistance should be firm and not spongy.
- There should be no signs of any fluid leaks anywhere in the braking system.
- The wheels should turn freely when the brakes are not applied.
- When the brakes are applied, the brakes on all four wheels must work, and the car must stop evenly in a straight line without pulling to one side.
- All the braking system components must be secure, and there must be no signs of excessive wear or corrosion.

PREPARING FOR THE MOT TEST

Tyres and wheels

- The tyres must be in good condition, and there must be no signs of excessive wear or damage (refer to 'Regular checks' on page 76).
- Tyres at the same end of the car must be of the same size and type. Always fit radial tyres; crossply tyres are not suitable for modern cars.

Seatbelts

- The mountings must be secure, and must not be loose or excessively corroded.
- The seatbelt webbing should not be frayed or damaged (this must be checked along the full length of the belts).
- When a seatbelt is fastened, the locking mechanism should hold securely and should release when intended.
- In the case of inertia reel seatbelts, the retractors should work properly when the belts are released.

General

- The windscreen wipers and washers must work properly, and the wiper blades must clear the windscreen without smearing.
- The swept area of the windscreen must have a zone 290mm wide, centred on the steering wheel, free of impact damage greater than 10mm across; this zone must be not encroached by any stickers by more than 10mm. Similar restrictions apply to the remainder of the swept area. Here, the damage or obstruction must not exceed 40mm across.
- The horn must work properly, be loud enough to be heard by other road users and emit a constant tone. Two-tone horns, which alternate between the two frequencies (i.e. they warble), are illegal.
- The exhaust mountings must be secure, and the system itself must be free from leaks and serious corrosion.
- There must be no serious corrosion or damage to the vehicle structure. Corrosion of body panels will not necessarily cause a car to fail the test, but there should be no serious corrosion of any of the 'load bearing' structural components.
- Door handles and locks should work properly from both inside and outside; boot lids and tailgates must close securely; seats and spare tyres must be securely mounted; petrol caps must seal and fasten properly.
- Registration plates must be securely fitted, front and rear, clearly legible, not obscured by items such as a tow bar, with the letters and figures correctly formed and spaced. (Letters or figures formed or positioned so that they are likely to be mis-read, i.e. to personalise a number plate, are now illegal – the vehicle will be failed.)
- Vehicles must display a legible Vehicle Identification Number (VIN). This is either a plate secured to the body or chassis in the engine compartment or a number stamped or etched on the vehicle.
- The exhaust gas emissions level must be within certain limits depending on the age of the car (special equipment is required to measure exhaust gas emissions), and the MOT test also includes a visual check for excessive exhaust smoke when the engine is running. Generally, the car should meet the emission regulations if regular servicing has been carried out.

VAUXHALL NOVA

FAULT FINDING

The following table isn't intended as an exhaustive guide to fault finding, but it summarises some of the more common faults which may crop up during a car's life.

Hopefully, the table should help you to find the cause of a problem, even if you can't cure it yourself.

When confronted with a fault, try to think calmly and logically about the symptom(s), and you should be able to work out what the fault can't be! Check one item at a time, otherwise if you do clear the fault, you may not know what was causing it.

The commonest cause of difficulty is starting, especially in the Winter. Make sure that your battery is kept fully charged, and that the ignition components are in good condition, clean and dry (refer to *'Servicing'* on page 97).

Further details of fault diagnosis, along with comprehensive renewal procedures for most components can be found in our Owners Workshop Manual for your car (OWM 909).

SYMPTOM	POSSIBLE CAUSES
ENGINE	
Starter motor doesn't turn, and headlights don't come on	● Flat battery ● Loose, dirty or corroded battery connections ● Other electrical or wiring fault
Starter motor doesn't turn, and headlights dim	● Battery charge low ● Loose, dirty or corroded battery connections ● Faulty starter motor ● Other electrical or wiring fault ● Seized engine
Starter motor doesn't turn, and headlights are bright	● Loose or dirty starter motor connections ● Faulty starter motor or ignition switch ● Other electrical or wiring fault
Starter motor spins but doesn't turn engine	● Faulty starter motor ● Engine fault (ie worn or damaged flywheel ring gear)
Engine turns slowly but won't start	● Battery charge low ● Electrical or wiring fault ● Wrong grade of engine oil (refer to *'Service specifications'* on page 69)

VAUXHALL NOVA

FAULT FINDING

SYMPTOM	POSSIBLE CAUSES
ENGINE (continued)	
Engine turns but won't fire, or engine starts but won't keep running	• Fuel tank empty • Damp or dirty ignition system components • Dirty or loose ignition system connections • Incorrectly-adjusted or faulty spark plugs (refer to *'Servicing'* on page 101) • Fuel system fault • Other electrical or wiring fault
Engine idles, but stalls when accelerator pedal is depressed	• Blocked or dirty air cleaner filter element (refer to *'Servicing'* on page 100) • Fuel system fault • Ignition system fault • Electrical fault
Poor acceleration, misfiring or lack of power	• Incorrectly-adjusted or faulty spark plugs (refer to *'Servicing'* on page 101) • Incorrect engine valve clearances (OHV engine only) • Fuel system fault • Ignition system fault • Electrical fault
Engine continues to run when ignition is switched off	• Engine overheating (possibly due to low coolant level – refer to *'Regular checks'* on page 75) • Incorrectly-adjusted spark plugs, or wrong type of spark plugs fitted (refer to *'Servicing'* on page 101) • Wrong grade of petrol • Fuel system fault • Excessive build-up of carbon inside engine ('decoke' required) • Faulty ignition switch • Other electrical or wiring fault
Engine doesn't reach normal operating temperature	• Faulty cooling system thermostat • Faulty temperature gauge or sensor

VAUXHALL NOVA

FAULT FINDING

SYMPTOM	POSSIBLE CAUSES

ENGINE (continued)

Engine overheats, or temperature gauge reads too high
- Airflow to radiator obstructed
- Blocked radiator, hose or engine coolant passage
- Coolant or engine oil level low (refer to *'Regular checks'* on page 74 or 75)
- Coolant hose(s) leaking, deteriorated or damaged
- Faulty cooling system thermostat
- Faulty water pump
- Faulty water pump drive (OHV engine)
- Faulty temperature gauge or sender
- Faulty electric cooling fan or switch

Ignition warning light comes on when engine is running
- Alternator drivebelt loose or broken (refer to *'Servicing'* on page 98)
- Faulty alternator
- Other electrical or wiring fault

Oil pressure warning light comes on when engine is running
- Oil level below 'MIN' mark on dipstick – refer to *'Regular checks'* on page 74 (if oil level is correct, and light is still on, seek advice, but **don't** start the engine)
- Oil leak
- Faulty oil pressure switch
- Badly-worn engine components

GEARBOX, TRANSMISSION AND CLUTCH

Difficulty in engaging gear (manual gearbox)
- Worn or faulty gearbox components
- Worn or faulty clutch mechanism
- Engine idle speed too high

Clutch slips (car does not accelerate when engine revs increase – manual gearbox)
- Incorrect clutch cable adjustment
- Faulty clutch mechanism
- Contaminated or worn clutch

FAULT FINDING

SYMPTOM	POSSIBLE CAUSES
BRAKES	
Brakes feel 'spongy'	• Air in brake hydraulic system • Fluid leak in brake system
Excessive brake pedal travel	• Faulty rear brake mechanism • Air in brake hydraulic system • Fluid leak in brake system
Brakes require excessive pedal pressure	• Damp, dirty or contaminated brake components • Fluid leak in brake system • Faulty or seized brake components • Faulty brake servo
Car pulls to one side	• Incorrect tyre pressure(s) or uneven tyre wear (refer to *'Regular checks'* on page 76) • Faulty or seized brake components • Worn or contaminated brake friction material • Incorrect wheel alignment • Worn or faulty steering components • Worn or faulty suspension components
Brakes squeal and/or judder	• Badly-worn or corroded brake components • Incorrectly-assembled brake components • Contaminated brake friction material • Worn or faulty suspension components • Worn or faulty steering components
Brake fluid level warning light comes on	• Low brake fluid level (refer to *'Regular checks'* on page 76) • Faulty fluid level sensor or wiring

VAUXHALL NOVA

FAULT FINDING

SYMPTOM | POSSIBLE CAUSES

SUSPENSION AND STEERING

Car becomes heavy to steer
- Low tyre pressure(s) (refer to *'Regular checks'* on page 76)
- Incorrect wheel alignment
- Worn or faulty steering components
- Worn or faulty suspension components

Car pulls to one side
- Refer to *'Brakes'* section of table

Car wanders
- Incorrect tyre pressure(s) or uneven tyre wear (refer to *'Regular checks'* on page 76)
- Car unevenly loaded
- Incorrect wheel alignment
- Worn or faulty suspension components
- Worn or faulty steering components

Car vibrates when driving
- Loose wheel bolt(s)
- Wheel(s) out of balance
- Worn or damaged driveshaft
- Worn or faulty suspension components
- Worn or faulty steering components
- Worn or faulty brake components

Hard or choppy ride
- Incorrect tyre pressure(s) (refer to *'Regular checks'* on page 76)
- Worn or faulty suspension components

Car leans excessively when cornering
- Roof rack overloaded
- Car unevenly loaded
- Worn or faulty suspension components

Uneven tyre wear
- Incorrect tyre pressure(s) (refer to *'Regular checks'* on page 76)
- Incorrect wheel alignment
- Wheel(s) out of balance
- Worn or faulty suspension components
- Worn or faulty steering components
- Brakes 'grabbing'

FAULT FINDING

SYMPTOM	POSSIBLE CAUSES
ELECTRICS	
Electrical systems don't work	• Loose, dirty or corroded battery connections • Faulty earth connection • Battery charge low • Blown fuse (refer to *'Bulb, fuse and relay renewal'* on page 118) • Faulty relay • Fuse link, connecting main wiring loom to battery blown (seek advice)
Direction indicators don't work	• Blown fuse (refer to *'Bulb, fuse and relay renewal'* on page 118) • Faulty earth connection • Faulty direction indicator relay • Faulty direction indicator switch
Bulbs burn out repeatedly	• Faulty connections at light socket • Faulty alternator regulator
All lights dim when engine speed drops to idle	• Loose alternator drivebelt (refer to *'Servicing'* on page 98) • Battery charge low • Faulty alternator

CAR JARGON

This Section should help you to understand some of the odd words of phrases used at garages and by 'car enthusiasts' which you may not be familiar with.

A

ABS – Abbreviation for Anti-lock Braking System. Uses sensors at each wheel to sense when the wheels are about to lock, and releases the brakes to prevent locking. This process occurs many times per second, and allows the driver to maintain steering control when braking hard.
Accelerator pump – A device attached to many *carburettors* which provides a spurt of extra fuel to the carburettor fuel/air mixture when the accelerator pedal is suddenly pressed down.
Additives – Compounds which are added to petrol and oil to improve their quality and performance.
Advance and retard – A system for altering the *ignition timing*.
AF – An abbreviation of 'Across Flats', the way in which many nuts, bolts and spanners are identified. AF is usually preceded by an Imperial unit of measurement – eg ½in AF. Unless otherwise stated, all metric measurements are assumed to be AF, so the abbreviation is not normally used for metric nuts, bolts and spanners.
Air cooling – Alternative method of engine cooling in which no water is used. An engine-driven fan forces air at high speed over the surfaces of the engine.
ALB – Abbreviation for Anti-Lock Braking System. See *'ABS'*.
Alternator – A device for converting rotating mechanical energy into electrical energy. In modern cars, it has superseded the dynamo for charging the battery because of its much greater efficiency.
Ammeter – A device for measuring electrical current – the current supplied to the battery by the alternator, or drawn from the battery by the car's electrical systems.
Antifreeze – A chemical mixed with the water in the cooling system to lower the temperature at which the coolant freezes, and in modern cars to prevent corrosion of the metal in the cooling system.
Anti-roll bar – A metal bar mounted transversely across the car, connecting the two sides of the suspension, which counteracts the natural tendency for the car to lean when cornering.
Aquaplaning – A word used to describe the action of a tyre skating across the surface of water.
Automatic transmission – A type of *gearbox* which selects the correct gear ratio automatically according to engine speed and load.
Axle – Spindle on which a wheel revolves.

B

Balljoint – A ball-and-socket type joint used in steering and *suspension* systems, which allows relative movement in more than one plane.
Battery condition indicator – A device for measuring electrical voltage (a voltmeter) connected via the ignition switch to the car battery. Unlike an *ammeter*, it gives an indication that a battery is close to failing. **Also**, most 'maintenance-free' batteries have a battery condition indicator fitted to their casing, which consists of a small disc which changes its colour when the battery is close to failure and requires renewal.
Bearing – Metal or other hard wearing surface against which another part moves, and which is designed to reduce friction and wear (bearings are usually lubricated).

VAUXHALL NOVA

CAR JARGON

Bendix drive – A device on some types of starter motor which allows the motor to drive the engine for starting, then disengages when the engine starts to run.

BHP – see *Horsepower*.

Big end – The end of a *connecting rod* which is attached to the *crankshaft*. It incorporates a *bearing* and transmits the linear movement of the connecting rod to the crankshaft.

Bleed nipple (or valve) – A hollow screw which allows air or fluid to be bled out of a system when it is loosened.

Brake caliper – The part of a *disc brake* system which houses the *brake pads* and the hydraulically-operated pistons.

Brake disc – A rotating disc, coupled to a roadwheel, which is clamped between hydraulically operated friction pads in a *disc brake* system.

Brake fade – A temporary loss of braking efficiency due to overheating of the brake friction material.

Brake pad – The part of a *disc brake* system which consists of the friction material and a metal backing plate.

Brake shoe – The part of a *drum brake* system which consists of the friction material and a curved metal former.

Breather – A device which allows fresh air into a system or allows contaminated air out.

Bucket tappet – A bucket shaped component used in some engines to transfer the rotary movement of the *camshaft* to the up-and-down movement required for *valve* operation.

Bump stop – A hard piece of rubber used in many *suspension* systems to prevent the moving parts from contacting the body during violent suspension movements.

C

Camber angle – The angle at which the front wheels are set from the vertical, when viewed from the front of the car. Positive camber is the amount in degrees which the wheels are tilted out at the top.

Cam follower – A piece of metal used to transfer the rotary movement of the *camshaft* to the up-and-down movement required for *valve* operation.

Camshaft – A rotating shaft driven from the *crankshaft* with lobes or cams used to operate the engine *valves* via the *valve gear*.

Carbon leads – Ignition HT leads incorporating carbon (black fibres) which eliminates the need for separate radio and TV *suppressors*.

Carburettor – A device which is used to mix air and fuel in the proportions required for burning by the engine under all conditions of engine running.

Castor angle – The angle between the front wheels pivot points and a vertical line when viewed from the side of the car. Positive castor is when the axis is inclined rearwards.

Catalytic converter – A device incorporated in the exhaust system which speeds up the natural decomposition of the exhaust gases, and reduces the amount of harmful gases released into the atmosphere. Cars fitted with catalytic converters must be operated on *unleaded petrol*, as *leaded petrol* will destroy the catalyst.

Centrifugal advance – System of ignition *advance and retard* incorporated in many *distributors* in which weights rotating on a shaft alter the ignition timing according to engine speed.

Choke – This has two common meanings. It is used to describe the device which shuts off some of the air in a *carburettor* during cold starting (in order to provide extra fuel), and it may be manually or automatically operated. It's also used as a general term to describe a carburettor throttle bore.

Clutch – A friction device which allows two rotating components to be coupled together smoothly, without the need for either rotating component to stop moving.

Coil spring – A spiral coil of spring steel used in many *suspension* systems.

Combustion chamber – Shaped area in the *cylinder head* into which the fuel/air mixture is compressed by the *piston* and where the spark from the *spark plug* ignites the mixture.

Compression ratio (CR) – A term used to describe the amount by which the fuel/air mixture is compressed as a *piston* moves from the bottom to the top of its travel, and expressed as a number. For example an 8.5:1 compression ratio means that the volume of fuel/air mixture above the piston when the piston is at the bottom of its stroke is 8.5 times

CAR JARGON

that when the piston is at the top of its stroke.
Compression tester – A special type of pressure gauge which can be screwed into a *spark plug* hole, which measures the pressure in the cylinder when the engine is turning but not firing. This gives an indication of engine wear or possible leaks.
Condenser (capacitor) – A device in a *contact breaker point distributor* which stores electrical energy and prevents excessive sparking at the contact breaker points.
Connecting rod ('con-rod') – Metal rod in the engine connecting a *piston* to the *crankshaft*.
Constant velocity (CV) joint – A joint used in *driveshafts*, where the instantaneous speed of the input shaft is exactly the same as the instantaneous speed of the output shaft at any angle of rotation. This does not occur in ordinary *universal joints*.
Contact breaker points – A device in the *distributor* which consists of two electrical points (or contacts) and a cam which opens and closes them to operate the *HT* electrical circuit which provides the spark at the *spark plugs*.
Crankcase – The area of the *cylinder block* below the *pistons* which houses the *crankshaft*.
Crankshaft – A cranked shaft which is driven by the *pistons* and provides the engine output to the *transmission*.
Crossflow cylinder head – A *cylinder head* in which the inlet and exhaust *valves* and *manifolds* are on opposite sides.
Crossply tyre – A tyre whose construction is such that the weave of the fabric material layers is running diagonally in alternately opposite directions to a line around the circumference of the tyre.
Cubic capacity – The total volume within the *cylinders* of an engine which is swept by the movement of the *pistons*.
CVH – A term applied by the Ford Motor Company to their overhead camshaft engines which incorporate a hemispherical *combustion chamber*. CVH means Compound Valve angle, Hemispherical combustion chamber.
Cylinder – Close fitting metal tube in which a *piston* slides. In the case of an engine, the cylinders may be bored directly into the *cylinder block*, or on some engines, cylinder liners are used which rest in the cylinder block and can be replaced when worn with matching pistons to avoid the requirement for *reboring* the cylinder block.
Cylinder block – The main engine casting which contains the *cylinders*, *crankshaft* and *pistons*.
Cylinder head – The casting at the top of the engine which contains the *valves* and associated operating components.

D

Damper – See *shock absorber*.
Dashpot – An oil-filled *cylinder* and *piston* used as a damping device in SU and Zenith/Stromberg CD type carburettors.
Dead axle (beam axle) – The simplest form of axle, consisting of a horizontal member attached to the car underbody by springs. This arrangement is used for the rear axle on some front-wheel-drive cars.
Decarbonising ('decoking') – Removal of all carbon deposits from the *combustion chambers* and the tops of the *pistons* and *cylinders* in an engine.
De Dion axle – A rear axle consisting of a cranked tube attached to the wheel hubs, with a separately mounted *differential* gear and *driveshafts*. *Suspension* is normally through *coil springs* between the wheel hubs and car underbody.
Derv – Abbreviation for Diesel-Engined Road Vehicle. A term often used to refer to Diesel fuel.
Diaphragm – A flexible membrane used in some components such as fuel pumps. The diaphragm spring used on *clutches* is similar, but is made from spring steel.
Diesel engine – An engine which relies on the heat generated when compressing air to ignite the fuel, and which therefore doesn't need an *ignition system*. Diesel engines have much higher *compression ratios* than petrol engines, normally around 20:1.
Differential – A system of gears (generally known as a crownwheel and pinion) which allows the *torque* provided by the engine to be applied to both driving wheels. The differential divides the torque proportionally between the driving wheels to allow one wheel to turn faster than the other, for example during cornering.

CAR JARGON

DIN – This stands for Deutsche Industrie Norm (German Industry Standard), which provides international standards for measuring engine power, torque, etc.

Disc brake – A brake assembly where a rotating disc is clamped between hydraulically operated friction pads.

Distributor – A device used to distribute the *HT* current to the individual *spark plugs*. The distributor may also contain the *advance and retard* mechanism. On some older cars, the distributor also contains the *contact breaker points* assembly.

Distributor cap – Plastic cap which fits on top of the *distributor* and contains electrodes, in which the *rotor arm* rotates to distribute the *HT* spark voltage to the correct *spark plug*.

DOHC – Abbreviation for Double Overhead Camshaft (see *'Twin-cam'*).

Doughnut – A term used to describe the flexible rubber coupling used in some *driveshafts*.

Driveshaft – Term usually used to describe the shaft (normally incorporating *universal* or *constant velocity joints*), which transmits drive from a *differential* to one wheel. More commonly found in front-wheel-drive cars.

Drive train – A collective term used to describe the *clutch/gearbox/transmission* and the other components used to transmit drive to the wheels.

Drum brake – A brake assembly with friction linings on 'shoes' running inside a cylindrical drum attached to the wheel.

Dual circuit brakes – A *hydraulic* braking system consisting of two separate fluid circuits, so that if one circuit becomes inoperative, braking power is still available from the other circuit.

Dwell angle – A measurement which corresponds to the number of degrees of *distributor* shaft rotation during which the *contact breaker points* are closed during the ignition cycle of one *cylinder*. The angle is altered by adjusting the contact breaker points gap.

E

Earth strap – A flexible electrical connection between the battery and a car earth point, or between the engine/*gearbox* and the car body to provide a return current path flow to the battery.

EFI – Abbreviation for Electronic *Fuel Injection*.

Electrode – An electrical terminal, eg in a *spark plug* or *distributor cap*.

Electrolyte – A current-conducting solution inside the battery (consisting of water and sulphuric acid in the case of a car battery).

Electronic ignition – An *ignition system* incorporating electronic components in place of *contact breaker points*, which can produce a much higher spark voltage than a contact breaker system, and is less affected by worn components.

Emission control – The reduction or prevention of the release into the atmosphere of poisonous fumes and gases from the engine and fuel system of a car. Required to different degrees by the laws of some countries, and achieved by engine design and the use of special devices and systems.

Epicyclic gears (planetary gears) – A gear system used in many *automatic transmissions* where there is a central 'sun' gear around which smaller 'planet' gears rotate.

Exhaust gas analyser – An instrument used to measure the amount of pollutants (mainly carbon monoxide) in a car's exhaust gases.

Expansion tank – A container used in many cooling systems to collect the overflow from the car's cooling system as the coolant heats up and expands.

F

Filter – A device for removing foreign particles from air, fuel or oil.

Final drive – A collective term (often expressed as a gear ratio) for the crownwheel and pinion (see *Differential*).

Flat engine – Form of engine design in which the *cylinders* are opposed horizontally, usually with an equal number on each side of the central *crankshaft*.

Float chamber – The part of a *carburettor* which contains a float and *needle valve* for

CAR JARGON

controlling the fuel level in the reservoir.
Flywheel – A heavy rotating metal disc attached to the *crankshaft* and used to smooth out the pulsing from the *pistons*.
Four stroke (cycle) – A term used to describe the four operating strokes of a *piston* in a conventional car engine. These are (1) Induction – drawing the air/fuel mixture into the engine as the piston moves down; (2) Compression – of the fuel/air mixture as the piston rises; (3) Power stroke – where the piston is forced down after the fuel/air mixture has been ignited by the *spark plug*, and (4) Exhaust stroke – where the piston rises and pushes the burnt gases out of the *cylinder*. During these operations, the inlet and exhaust *valves* are opened and closed at the correct moment to allow the fuel/air mixture in, the exhaust gases out, or to provide a gas-tight compression chamber.
Fuel injection – A method of injecting fuel into an engine. Used in *Diesel engines* and also on some petrol engines in place of a *carburettor*.
Fuel injector – Device used on *fuel injection* engines to inject fuel directly or indirectly into the *combustion chamber*. Some fuel injection systems use a single fuel injector, while some systems use one fuel injector for each *cylinder* of the engine.

G

Gasket – Compressible material used between two surfaces to provide a leakproof joint.
Gearbox – A group of gears and shafts installed in a housing, positioned between the *clutch* and the *differential*, and used to keep the engine within its safe operating speed range as the speed of the car changes.

H

Half-shaft – A *driveshaft* used to transmit the drive from the *differential* to one of the rear wheels.
Hardy-Spicer joint (Hooke's or Cardan joint) – See *Universal joint*.
Helical gears – Gears in which the teeth are cut at an angle across the circumference of the gear to give a smoother mesh between gears and quieter running.
Horsepower – A measurement of power. Brake Horsepower (BHP) is a measure of the power required to stop a moving body.
HT – Abbreviation of High Tension (meaning high voltage) used to describe the *spark plug* voltage in an *ignition system*.
Hub carrier – A component usually found at each front corner of a car which carries the wheel and brake assembly, and to which the *suspension* and steering components are attached.
Hydraulic – A term used to describe the operation of a system by means of fluid pressure.
Hypoid gear – A gear with curved teeth which transmits drive through a right-angle, where the centreline of the drive gear is offset from the centreline of the driven gear. The meshing action of hypoid gears allows a larger and therefore stronger drive gear, and the meshing noise is reduced in comparison with conventional gears.

I

Independent suspension – A *suspension* system where movement of one wheel has no effect on the movement of the other, eg independent front suspension.
Ignition coil – An electrical coil which forms part of the *ignition system* and which generates the *HT* voltage.
Ignition system – The electrical system which provides the spark to ignite the air/fuel mixture in the engine. Normally the system consists of the battery, *ignition coil*, *distributor*, ignition switch, *spark plugs* and wiring.
Ignition timing – The time in the *cylinder* firing cycle at which the ignition spark (provided by the *spark plug*) occurs. The spark timing is normally a few degrees of *crankshaft* rotation before the *piston* reaches the top of its stroke, and is expressed as a number of degrees before top-dead-centre (BTDC).
Inertia reel – Automatic type of seat belt mechanism which allows the wearer to move freely in normal use, but which locks on sensing either sudden deceleration or a sudden movement of the wearer.

CAR JARGON

In-line engine – An engine in which the *cylinders* are positioned in one row as opposed to being in a *flat* or *vee* configuration.

J

Jet – A calibrated nozzle or orifice in a *carburettor* through which fuel is drawn for mixing with air.
Jump leads – Heavy electric cables fitted with clips to enable a car's battery to be connected to another battery for emergency starting.

K

Kerb weight – The weight of a car, unladen but ready to be driven, ie with enough fuel, oil, etc, to travel an arbitrary distance.
Kickdown – A device used on *automatic transmissions* which allows a lower gear to be selected for improved acceleration by fully depressing the accelerator.
Kingpin – A device which allows the front wheel of a car to swivel about a near vertical axis.
Knocking – See *'Pinking'*.

L

Laminated windscreen – A windscreen which has a thin plastic layer sandwiched between two layers of toughened glass. It will not shatter or craze when hit.
Lead-free petrol – Contains no lead. It has no lead added during manufacture, and the natural lead content is refined out. This type of petrol is not currently available for general use in the UK, and should not be confused with *unleaded* petrol.
Leaded petrol – Normal 4-star petrol. Has a low amount of lead added during manufacture, in addition to the natural lead found in crude oil.
Leading shoe – A *drum brake* shoe of which the leading end (the one moved by the operating *pistons*) is reached first by a given point on the drum during normal forward rotation. A simple drum brake will have one leading and one trailing (the opposite) shoe.

Leaf spring – A spring commonly used on cars with a *live axle*, consisting of several long curved steel plates clamped together.
Limited slip differential – A type of *differential* which prevents one wheel from standing still while the other wheel spins excessively. Often used on high-performance cars.
Live axle – An axle through which power is transmitted to the rear wheels.
Loom – A complete car wiring system or section of a wiring system consisting of all the wires of correct length, etc, to wire up the various circuits.
LT – Abbreviation of Low Tension (meaning low voltage), used to describe battery voltage in the *ignition system*.

M

MacPherson strut – An independent front *suspension* system where the swivelling, springing and shock absorbing action of the wheels is dealt with by a single assembly.
Manifold – A device used for ducting the air/fuel mixture to the engine (inlet manifold), or the exhaust gases from the engine (exhaust manifold).
Master cylinder – A *cylinder* containing a *piston* and *hydraulic* fluid, directly coupled to a foot pedal (eg brake or *clutch* master cylinder). Used for transmitting pressure to the brake or clutch operating mechanism.
Metallic paint – Paint finish incorporating minute particles of metal to give added lustre to the colour.
Multigrade – Lubricating oil whose *viscosity* covers that of several single grade oils, making it suitable for use over a wider range of operating conditions.

N

Needle bearing – Type of *bearing* in which needle or cone-shaped rollers are used around the circumference to reduce friction.
Needle valve – A component of the *carburettor* which restricts the flow of fuel or fuel/air mixture according to the position of the valve in an orifice or *jet*.

CAR JARGON

Negative earth – Electrical system (almost universally adopted) in which the negative terminal of the car battery is connected to the car body. The polarity of all the electrical equipment is determined by this.

O

Octane rating – A scale rating for grading petrol.
ohc (overhead cam) – Describes an engine in which the *camshaft* is situated above the *cylinder head*, and operates the *valve gear* directly.
ohv (overhead valve) – Describes an engine which has its *valves* in the *cylinder head*, but with the *valve gear* operated by *pushrods* from a *camshaft* situated lower in the engine.
Oil cooler – Small *radiator* fitted in the oil circuit and positioned in a cooling airflow to cool the oil. Used mainly on high-performance engines.
Overdrive – A device coupled to a car's *gearbox* which raises the output gear ratio above the normal 1:1 of top gear.
Oversteer – A tendency for a car to turn more tightly into a corner than intended.

P

PCV (Positive Crankcase Ventilation) – A system which allows fumes and vapours which build up in the *crankcase* to be drawn into the engine for burning.
Pinion – A gear with a small number of teeth which meshes with one having a larger number of teeth.
Pinking – A metallic noise from the engine often caused by the *ignition timing* being too far *advanced*. The noise is the result of pressure waves which cause the *cylinder* walls to vibrate when the ignited fuel/air mixture is compressed.
Piston – Cylindrical component which slides in a closely-fitting metal tube or *cylinder* and transmits pressure. The pistons in an engine, for example, compress the fuel/air mixture, transmit the power to the *crankshaft*, and push the burnt gases out through the exhaust *valves*.

Piston ring – Hardened metal ring which is a spring fit in a groove running round the *piston* to ensure a gas-tight seal between the piston and *cylinder* wall.
Positive earth – The opposite of *negative earth*.
Power steering – A steering system which uses *hydraulic* fluid pressure (provided by an engine-driven pump) to reduce the effort required to steer the car.
Pre-ignition – See *'Pinking'*.
Propeller shaft – The shaft which transmits the drive from the *gearbox* to the rear axle in a front-engined rear-wheel-drive car.
Pushrod – A rod which is moved up and down by the rotary motion of the *camshaft* and operates the *rocker arms* in an *ohv* engine.

Q

Quarter light – A triangular window mounted in front or behind the main front or rear windows, usually in the front door, or behind the rear door.
Quartz-halogen bulb – A bulb with a quartz envelope (instead of glass), filled with a halogen gas. Gives a brighter, more even spread of light than an ordinary bulb.

R

Rack and pinion – Simplest form of steering mechanism which uses a *pinion* gear to move a toothed rack.
Radial ply tyre – A tyre in which the fabric material plies are arranged laterally, at right angles to the circumference.
Radiator – Cooling device through which the engine coolant is passed, situated in an air flow and consisting of a system of fine tubes and fins for rapid heat dissipation.
Radius arms (rods) – Locating arms sometimes used with a *live axle* to positively locate it in the fore-and-aft direction.
Rebore – The process of enlarging the *cylinder* bores to a very accurately specified measurement in order to fit new *pistons* to overcome wear in the engine. Not normally necessary unless the engine has covered a very high mileage.

CAR JARGON

Recirculating ball steering – A derivative of *worm and nut* steering, where the steering shaft motion is transmitted to the steering linkage by balls running in the groove of a worm gear.
Rev counter – See *Tachometer*.
Rocker arm – A lever which rocks on a central pivot, with one end moved up and down by the *camshaft*, and the other end operating an inlet or exhaust *valve*.
Rotary engine – See '*Wankel engine*'.
Rotor arm – A rotating arm in the *distributor* which distributes the *HT* spark voltage to the correct *spark plug*.
Running on – A tendency for an engine to keep on running after the ignition has been switched off. Often caused by a badly maintained engine or the use of an incorrect grade of fuel.

S

SAE – Society of Automotive Engineers (of America). Lays down international standards for the classification of engine performance and many other specifications, but is most commonly used to classify oils.
Safety rim – A special wheel rim shape which prevents a deflated tyre from rolling off the wheel.
Sealed beam – A sealed headlamp unit where the filament is an integral part and cannot be renewed separately.
Semi-trailing arm – A common form of independent rear *suspension*.
Servo – A device for increasing the normal effort applied to a control.
Shock absorber – A device for damping out the up-and-down movement of the *suspension* when the car hits a bump in the road.
Spark plug – A device with two *electrodes* insulated from each other by a ceramic material, which screws into an engine *combustion chamber*. When the *HT* voltage is applied to the plug terminal, a spark jumps across the electrodes and ignites the fuel/air mixture.
Squab – Another name for a seat cushion.
Steel-braced tyre – Tyre in which extra plies containing steel cords are incorporated with the fabric plies to give added strength.

Steering arm (knuckle) – A short arm on the front *hub carrier* to which the steering linkage connects.
Steering gear – A general term used to describe the steering components, usually refers to a steering rack-and-pinion assembly.
Steering rack – See *Rack and pinion*.
Stroboscopic light – A light switched on and off by the engine *ignition system* which is used for checking the *ignition timing* when the engine is running.
Stroke – The total distance travelled by a single *piston* in its *cylinder*.
Stub axle – A short axle which carries one wheel.
Subframe – A small frame which is mounted on the car's body, and carries the *suspension* and/or the *drivetrain* assemblies.
Sump – The main reservoir for the engine oil.
Supercharger – A device which uses an engine-driven turbine (usually driven by a belt or gears from the *crankshaft*) to drive a compressor which forces air into the engine, providing increased fuel/air mixture flow, and therefore increased engine efficiency. Sometimes used on high-performance engines.
Suppressor – A device which is used to reduce or eliminate electrical interference caused by the *ignition system* or other electrical components.
Suspension – A general term used to describe the components which suspend the car body on its wheels.
Swing axle – A *suspension* arm which is pivoted near the front-to-rear centreline of the car, and which allows the wheel to swing vertically about that pivot point.
Synchromesh – A device in a *gearbox* which synchronises the speed of one gear shaft with another to produce smooth, noiseless engagement of the gears.

T

Tachometer – Also known as a rev counter, indicates engine speed in revolutions per minute (rpm).
Tappet – A term often used to refer to the component which transmits the rotary *camshaft* movement to the up-and-down movement required for *valve* operation.

VAUXHALL NOVA

CAR JARGON

Thermostat – A device which is sensitive to changes in engine coolant temperature, and opens up an additional path for coolant to flow through the *radiator* (to increase the cooling) when the engine has warmed up.

Tie-rod – A rod which connects the *steering arms* to the *steering gear*.

Timing belt – Fabric or rubber belt engaging on sprocket wheels and driving the *camshaft* from the *crankshaft*.

Timing chain – Metal flexible link chain engaging on sprocket wheels and driving the *camshaft* from the *crankshaft*.

Timing marks – Marks normally found on the *crankshaft* pulley or the *flywheel* and used for setting the ignition firing point with respect to a particular *piston*.

Toe-in/toe-out – The amount by which the front wheels point inwards or outwards from the straight-ahead position when steering straight ahead.

Top Dead Centre (TDC) – The point at which a *piston* is at the top of its *stroke*.

Torque – The turning force generated by a rotating component.

Torque converter – A coupling where the driving *torque* is transmitted through oil. At low speeds there is very little transfer of torque from the input to the output. As the speed of the input shaft increases, the direction of fluid flow within a system of vanes changes, and torque from the input impeller is transferred to the output turbine. The higher the input speed, the closer the output speed approaches it, until they are virtually the same.

Torsion bar – A metal bar which twists about its own axis, and is used in some *suspension* systems.

Toughened windscreen – A windscreen which when hit, will shatter in a particular way to produce blunt-edged fragments or will craze over but remain intact. A zone toughened windscreen has a zone in front of the driver which crazes into larger parts to reduce the loss of visibility which occurs when toughened windscreens break, but is otherwise similar.

Track rod – See *Tie-rod*.

Trailing arm – A form of independent *suspension* where the wheel is attached to a swinging arm, and is mounted to the rear of the arm pivot.

Transaxle – A combined *gearbox*/axle assembly from which two *driveshafts* transmit the drive to the wheels.

Transmission – A general term used to describe some or all of the *drivetrain* components excluding the engine, most commonly used to describe automatic gearboxes.

Turbocharger – A device which uses a turbine driven by the engine exhaust gases to drive a compressor which forces air into the engine, providing increased fuel/air mixture flow, and therefore increased engine efficiency. Commonly used on high-performance engines.

Twin-cam – Abbreviation for twin overhead *camshafts* (see *'ohc'*). Used on engines with a *crossflow cylinder head*, usually with one camshaft operating the inlet *valves* and the other operating the exhaust valves. Gives improved engine efficiency due to improved fuel/air mixture and exhaust gas flow in the *combustion chambers*.

Two stroke (cycle) – A common term used to describe the operation of an engine where each downward *piston* stroke is a power stroke. The fuel/air mixture is directed to the crankcase where it's compressed by the descending piston and pumped into the *combustion chamber*. As the piston rises, the mixture is compressed and ignited, which forces the piston down. The burnt gases flow from the exhaust port, but the piston is now compressing another fuel/air mixture charge in the crankcase which repeats the cycle.

U

Understeer – A tendency for a car to go straight on when turned into a corner.

Universal joint – A joint that can swivel in any direction whilst at the same time transmitting *torque*. This type of joint is commonly used in *propeller shafts* and some *driveshafts*, but is not suitable for some applications because the input and output shaft speeds are not the same at all positions of angular rotation. The type in common use is known as a Hardy-Spicer, Hooke's or Cardan joint.

Unleaded petrol – Has no lead added during manufacture, but still has the natural lead content of crude oil. Generally available in the

CAR JARGON

UK, most modern cars can use this type of petrol, but seek advice first, as engine adjustments may be required. Engine damage can occur if unleaded petrol is used incorrectly. Not to be confused with *lead-free* petrol which is not currently available in the UK.
Unsprung weight – The part of the car which is not supported by the springs.

V

Vacuum advance – System of ignition *advance and retard* used in some *distributors* where the vacuum in the engine inlet *manifold* is used to act on a *diaphragm* which alters the *ignition timing* as the vacuum changes due to the throttle position.
Valve – A device which opens or closes to allow or stop gas or fluid flow.
Valve gear – A general term used for the components which are acted on by the *camshaft* in order to operate the *valves*.
16-valve – Term used to describe a four *cylinder* engine with four *valves* per cylinder (usually two inlet valves and two exhaust valves). Gives improved engine efficiency due to improved fuel/air mixture and exhaust gas flow in the *combustion chambers*.
Vee engine – An engine design in which the *cylinders* are set in two banks forming a 'V' when viewed from one end. A V8 for example consists of two rows of four cylinders each.
Venturi – A streamlined restriction in the *carburettor* throttle bore which causes a low pressure to occur; this sucks fuel into the air stream to form a vapour suitable for combustion.
Viscosity – A term used to describe the resistance of a fluid to flow. When associated with lubricating oil, it's given an *SAE* number, 10 being a very light oil and 140 being a very heavy oil.
Voltage regulator – A device which regulates the *alternator* output to a predetermined level. On most alternators the voltage regulator is an integral part of the alternator, and regulates the charging current as well as the voltage.

W

Wankel engine – A rotary engine which has a triangular shaped rotor which performs the function of the *pistons* in a conventional engine, and rotates in a housing shaped approximately like a broad-waisted figure of eight. Very few cars use this type of engine.
Wheel balancing – Adding small weights to the rim of a wheel so that there are no out-of-balance forces when the wheel rotates.
Wishbone – An 'A'-shaped *suspension* component, pivoted at the base of the 'A' and carrying a wheel at the apex. Normally mounted close to the horizontal.
Worm and nut steering – A steering system where the lower end of the steering column has a coarse screw thread on which a nut runs. The nut is attached to a spindle which carries the drop arm which, in turn, moves the steering linkage.

LOCAL RADIO FREQUENCIES

LOCAL RADIO

A comprehensive network of local radio stations now exists throughout the UK.

Most of these radio stations provide regularly updated reports on traffic flow and road conditions, which can be of great help to drivers. Listening to traffic reports will help you to avoid the inevitable 'jams' which occur during every day driving.

Some car radio/cassette players are now equipped with a 'Radio Data System' (RDS) which will automatically tune into special traffic information signals, interrupting normal radio or tape listening when bulletins are broadcast. An RDS-equipped radio may prove to be a worthwhile investment if you travel by car regularly, particularly when driving on business trips.

The following table provides details of all the local radio frequencies throughout the UK, region-by-region. Where stations broadcast on FM/VHF and AM/MW, it's suggested that the FM/VHF frequency is used wherever possible, as this will usually give better reception.

LOCAL RADIO FREQUENCIES

AREA	FM/VHF	AM/MW
AVON		
Bath		
BBC Bristol	104.6	1548
GWR FM	103.0	–
Bristol		
BBC Bristol	94.9	1548
Brunel Classic Gold	–	1260
Galaxy Radio	97.2	–
GWR FM	96.3	–
BEDFORDSHIRE		
Bedford		
BBC Bedfordshire	95.5	1161
Chiltern Radio	96.9	–
SuperGold	–	792
Luton		
BBC Bedfordshire	103.8	630
Chiltern Radio	97.6	–
SuperGold	–	828
BERKSHIRE		
Reading		
210 Classic Gold Radio	–	1431
210 FM	97.0	–
BBC Berkshire	104.4	–
Wokingham		
BBC Berkshire	104.1	–
BIRMINGHAM		
Birmingham		
BBC WM	95.6	1458
BRMB FM	96.4	–
Buzz FM	102.4	–
Xtra AM	–	1152
BORDERS		
Eyemouth		
Radio Borders	103.4	–
Peebles		
Radio Borders	103.1	–
Selkirk		
Radio Borders	96.8	–
BUCKINGHAMSHIRE		
Milton Keynes		
BBC Bedfordshire	104.5	630
Horizon Radio	103.3	–
CAMBRIDGESHIRE		
Cambridge		
BBC Cambridge	96.0	1026
CN FM	103.0	–
Peterborough		
BBC Cambridge	95.7	1447
Hereward Radio	102.7	1332
CHANNEL ISLANDS		
Guernsey		
BBC Guernsey	93.2	1116
Jersey		
BBC Jersey	88.8	1026
CHESHIRE		
Echo 96	96.4	–
Chester		
BBC Merseyside	95.8	1485
Congleton		
Signal Cheshire	104.9	–
Macclesfield		
BBC Stoke	94.6	1503
Warrington		
BBC Merseyside	95.8	1485
CLEVELAND		
Middlesbrough		
BBC Cleveland	95.0	1548
TFM	96.6	–
GNR	–	1170
CLWYD		
Prestatyn		
BBC Cymru/Wales	94.2	882
Rhyl		
BBC Cymru/Wales	94.2	882
Wrexham		
BBC Cymru/Wales	93.3	657
MFM 1034	97.1/103.4	–
Marcher Gold	–	1260
CORNWALL		
Isles of Scilly		
BBC Cornwall	96.0	–
Liskeard		
BBC Cornwall	95.2	657
Redruth		
BBC Cornwall	103.9	630
CUMBRIA		
Barrow-in-Furness		
BBC Furness	96.1	837
Kendal		
BBC Cumbria	95.2	–
Windermere		
BBC Cumbria	104.2	–
Workington		
BBC Cumbria	95.6	1458
DERBYSHIRE		
Chesterfield		
BBC Sheffield	94.7	1035
Derby		
BBC Derby	94.2	1116
Trent FM	102.8	–
Gem AM	–	945
Matlock		
BBC Derby	95.3	–
DEVON		
Barnstaple		
BBC Devon	94.8	801
Exeter		
BBC Devon	95.8	990
DevonAir	97.0	666/954
South West	103.0	–
Okehampton		
BBC Devon	96.0	801
Plymouth Area		
BBC Devon	103.4	855
Plymouth Sound	97.0	1152
Tavistock		
Radio in Tavistock	96.6	–
Torbay		
BBC Devon	103.4	1458
DevonAir	96.4	666/954
DORSET		
Poole		
BBC Solent	96.1	1359
Bournemouth		
BBC Solent	96.1	1359
2CR Classic Gold	–	828
2CR FM	102.3	–
Weymouth		
BBC Solent	96.1	–
DUMFRIES & GALLOWAY		
Dumfries		
BBC Scotland/Solway	94.7	585
SW Sound	97.2	–
Stranraer		
BBC Scotland	94.7	–
DURHAM		
Darlington		
BBC Newcastle	95.4	–
Durham		
BBC Newcastle	95.4	1458
DYFED		
Aberystwyth		
BBC Cymru/Wales	93.1	882

VAUXHALL NOVA

LOCAL RADIO FREQUENCIES   141

AREA	FM/VHF	AM/MW
EAST SUSSEX		
Brighton		
BBC Sussex	95.3	1485
South Coast Radio	–	1323
Southern Sound Classic Hits	103.5	–
Eastbourne		
BBC Sussex	104.5	1161
Southern Sound Classic Hits	102.4	–
Hastings		
Southern Sound Classic Hits	97.5	–
Newhaven		
Southern Sound Classic Hits	96.9	–
ESSEX		
Basildon		
BBC Essex	95.3	765
Chelmsford		
BBC Essex	103.5	765
Breeze AM		1431/1359
Essex Radio	102.6	–
Colchester		
BBC Essex	103.5	729
Mellow 1557	–	1557
Harlow		
BBC Essex	–	765
Manningtree		
BBC Suffolk	103.9	–
Southend-on-Sea		
BBC Essex	95.3	1530
Breeze AM		1431/1359
Essex Radio	96.3	–
FIFE		
Kirkcaldy		
BBC Scotland	94.3	810
GLOUCESTERSHIRE		
Gloucester		
BBC Gloucester	104.7	603
Severn Sound	102.4	
Three Counties Radio		774
Stroud		
BBC Gloucester	95.0	603
Severn Sound	103.0	–
GRAMPIAN		
Aberdeen		
BBC Scotland/Aberdeen	93.1	990
North Sound	96.9	1035
Stirling		
Central FM	96.7	–

AREA	FM/VHF	AM/MW
GWENT		
Newport		
Red Dragon Radio	97.4	–
Touch AM		1305
GWYNEDD		
Anglesey		
BBC Cymru/Wales	94.2	882
HAMPSHIRE		
Andover		
210 FM	102.9	–
Basingstoke		
210 FM	102.9	–
BBC Berkshire	104.1	–
Isle of Wight		
BBC Solent	96.1	999/1359
Isle of Wight Radio	–	1242
Portsmouth		
BBC Solent	96.1	999
Oceansound	97.5	–
South Coast Radio	–	1170
Southampton		
BBC Solent	96.1	999
Oceansound	97.5	–
Power FM	103.2	–
South Coast Radio	–	1557
Winchester		
Oceansound	96.7	–
Power FM	103.2	–
HEREFORDSHIRE		
Hereford		
BBC Here/Worc	94.7	819
Radio Wyvern	97.6	954
HERTFORDSHIRE		
Watford		
BBC GLR	94.9	1458
St Albans		
BBC Bedfordshire	103.8	630
Stevenage		
BBC Bedfordshire	103.8	630
HIGHLAND		
Inverness		
BBC Scotland	94.0	810
Moray Firth Radio	97.4	1107
Fort William		
BBC Scotland	93.7	–
Oban		
BBC Scotland	93.3	–
Thurso		
BBC Scotland	94.5	–

AREA	FM/VHF	AM/MW
HUMBERSIDE		
Kingston upon Hull Area		
BBC Humberside	95.9	1485
Classic Gold	–	1161
Viking FM	96.9	–
ISLE OF MAN		
Douglas		
Manx Radio	97.2	1368
Snaefell		
Manx Radio	89.0	–
KENT		
Ashford		
Invicta FM	96.1	–
Canterbury		
BBC Kent	104.2	774
Invicta FM	102.8	–
Dover		
BBC Kent	104.2	774
Invicta FM	97.0	–
East Kent		
Coast Classics	–	603
Folkestone		
BBC Kent	104.2	774
Maidstone & Medway		
BBC Kent	96.7	1035
Coast Classics	–	1242
Invicta FM	103.1	–
Royal Tunbridge Wells		
BBC Kent	96.7	1602
Thanet		
Invicta FM	95.9	–
LANCASHIRE		
Burnley Area		
BBC Lancashire	95.5	855
Blackpool		
BBC Lancashire	103.9	855
Red Rose Gold	–	999
Red Rose Rock FM	97.4	–
Lancaster		
BBC Lancashire	104.5	1557
Preston		
BBC Lancashire	103.9	855
Red Rose Gold	–	999
Red Rose Rock FM	97.4	–
LEICESTERSHIRE		
Leicester		
BBC Leicester	104.9	837
Gem AM	–	1260
Sound FM	103.2	–

VAUXHALL NOVA

LOCAL RADIO FREQUENCIES

AREA	FM/VHF	AM/MW
LINCOLNSHIRE		
Lincoln Area		
BBC Lincoln	94.9	1368
Lincs FM	102.2	–
LONDON		
Brixton		
Choice FM	96.9	–
Ealing		
Sunrise Radio	–	1413
Greater London		
BBC GLR	94.9	1458
Capital FM	95.8	–
Capital Gold	–	1548
Jazz FM	102.2	–
Kiss FM	100.0	–
LBC Newstalk	97.3	–
Melody Radio	104.9	–
Spectrum Int. Radio	–	558/990
Haringey		
London Greek Radio	103.3	–
WNK	103.3	–
Thamesmead		
London Talkback Radio	–	1152
RTM	103.8	–
Southall		
Sunrise Radio	–	1413
LOTHIAN		
Bathgate		
Radio Forth	97.3	–
Dunfermline		
BBC Scotland	94.3	810
Edinburgh		
BBC Scotland	94.3	810
Max AM	–	1548
Radio Forth	97.3/97.6	–
MANCHESTER		
Manchester		
BBC GMR	95.1	1458
KFM	104.9	–
Piccadilly Gold	–	1152
Piccadilly Key 103	103.0	–
Sunset Radio	102.0	–
MERSEYSIDE		
Liverpool		
BBC Mersey	95.8	1485
City Talk	–	1548
Radio City	96.7	–
MID GLAMORGAN		
Aberdare		
BBC Cymru/Wales	93.6	882
Merthyr Tydfil		
BBC Cymru/Wales	96.8	882

AREA	FM/VHF	AM/MW
MIDDLESEX		
Hounslow		
Sunrise Radio	–	1413
NORFOLK		
King's Lynn		
BBC Norfolk	104.4	873
Norwich Area		
BBC Norfolk	95.1	855
Broadland	102.4	1152
NORTHAMPTONSHIRE		
Kettering		
KCBC	–	1530
Northampton		
BBC Northampton	104.2	1107
Northants	96.6	–
SuperGold	–	1557
NORTHUMBERLAND		
Berwick-upon-Tweed		
BBC Newcastle	96.0	–
Radio Borders	97.5	–
NOTTINGHAM		
Nottingham		
BBC Nottingham	103.8	1521
Trent FM	96.2/96.5	–
Gem AM	–	999
ORKNEY		
Kirkwall		
BBC Scotland	93.7	–
OXFORDSHIRE		
Banbury		
Fox FM	97.4	–
Oxford		
BBC Oxford	95.2	1485
Fox FM	102.6	–
POWYS		
Welshpool		
BBC Cymru/Wales	94.0	882
SHETLAND		
Lerwick		
BBC Scotland	92.7	–
SIBC	96.2	–
SHROPSHIRE		
Ludlow		
BBC Shropshire	95.0	1584
RFM	97.6	–
Shrewsbury		
BBC Shropshire	96.0	–
Beacon Radio	97.2/103.1	–
WABC	–	1017
Telford		
BBC Shropshire	96.0	756
Beacon Radio	97.2/103.1	–
WABC	–	1017

AREA	FM/VHF	AM/MW
SOMERSET		
Mendip Area		
Orchard FM	102.6	–
Wells		
BBC Bristol	95.5	1548
SOUTH GLAMORGAN		
Cardiff		
BBC Cymru/Wales	96.8	882
Red Dragon Radio	103.2	–
Touch AM	–	1359
STAFFORDSHIRE		
Stafford		
Echo 96	96.9	–
Stoke on Trent		
BBC Stoke	94.6	1503
Signal Radio	102.6	1170
STRATHCLYDE		
Ayr		
West Sound	96.7	1035
Girvan		
West Sound	97.5	–
Glasgow		
BBC Scotland	94.3	810
Clyde 1	102.5	–
Clyde 2	–	1152
Radio Clyde FM	97.3	–
Greenock		
BBC Scotland	94.3	810
Kilmarnock		
BBC Scotland	93.9	810
SUFFOLK		
Bury St. Edmunds		
Saxon Radio	96.4	1251
Great Barton		
BBC Suffolk	104.6	–
Ipswich		
BBC Suffolk	103.9	–
Radio Orwell	97.1	1170

VAUXHALL NOVA

LOCAL RADIO FREQUENCIES 143

AREA	FM/VHF	AM/MW
SURREY		
Guildford		
BBC Surrey	104.6	–
Delta Radio	97.1	–
First Gold Radio	–	1476
Premier Radio	96.4	–
Reigate		
Radio Mercury	102.7	1521
TAYSIDE		
Dundee		
Radio Tay	102.8	1161
BBC Scotland	92.7	810
Perth		
Radio Tay	96.4	1584
TYNE & WEAR		
Fenham		
BBC Newcastle	104.4	–
Metro FM	103.0	–
Newcastle Area		
BBC Newcastle	95.4	1458
GNR	–	1152
Metro FM	97.1	–
Sunderland		
Wear FM	103.4	–
ULSTER		
Belfast		
BBC Ulster	94.5	1341
Classic Trax BCR	96.7	–
Cool FM	97.4	–
Downtown Radio	102.6	–
Enniskillen		
BBC Ulster	93.8	873
Downtown Radio	96.6	–
Limavady		
Downtown Radio	96.4	–
Londonderry		
BBC Foyle	93.1	792
Downtown Radio	102.4	–
WARWICKSHIRE		
Leamington Spa		
Mercia FM	102.9	–

AREA	FM/VHF	AM/MW
WEST GLAMORGAN		
Swansea		
BBC Cymru/Wales	93.9	882
Swansea Sound	96.4	1170
WEST MIDLANDS		
Coventry		
BBC CWR	94.8	–
Mercia FM	97.0	–
Radio Harmony	102.6	–
Xtra AM	–	1359
Sutton Coldfield		
BBC Derby	104.5	–
Beacon Radio	97.2/103.1	–
Wolverhampton		
BBC WM	95.6	828
Beacon Radio	97.2/103.1	–
WABC	–	990
WEST SUSSEX		
Crawley		
Radio Mercury	102.7	1521
Horsham		
BBC Sussex	95.1	1368
Worthing		
BBC Sussex	95.3	1485
WESTERN ISLES		
Stornoway (Lewis)		
BBC Scotland	94.2	–
WILTSHIRE		
Chippenham		
BBC Wiltshire Sound	104.3	–
Marlborough		
GWR FM	96.5	–
Salisbury		
BBC Wiltshire Sound	103.5	–
Swindon		
BBC Wiltshire Sound	103.6	1368
Brunel Classic Gold	–	1161
GWR FM	97.2	–
West Wilts		
Brunel Classic Gold	–	936
GWR FM	102.2	–
WORCESTERSHIRE		
Worcester		
BBC Here/Worc	104.0	738
Radio Wyvern	102.8	1530

AREA	FM/VHF	AM/MW
YORKSHIRE NORTH		
Northallerton		
BBC York	104.3	–
Scarborough		
BBC York	95.5	1260
Whitby		
BBC Cleveland	95.8	–
York		
BBC York	103.7	666
YORKSHIRE SOUTH		
Barnsley		
Classic Gold	–	1305
Hallam	102.9	–
Doncaster		
BBC Sheffield	104.1	1035
Classic Gold	–	990
Hallam	103.4	–
Rotherham		
BBC Sheffield	88.6	1035
Hallam	96.1	–
Sheffield		
BBC Sheffield	88.6	1035
Classic Gold	–	1548
Hallam	97.4	–
YORKSHIRE WEST		
Bradford		
Classic Gold	–	1278
Pennine	97.5	–
Sunrise FM	103.2	–
Halifax		
Classic Gold	–	1530
Pennine	102.5	–
Huddersfield		
Classic Gold	–	1530
Pennine	102.5	–
Leeds		
Aire FM	96.3	–
BBC Leeds	92.4	774
Magic 828	–	828

VAUXHALL NOVA

CONVERSION FACTORS

Length (distance)

Inches (in)	x 25.4	= Millimetres (mm)	x 0.0394	= Inches (in)
Feet (ft)	x 0.305	= Metres (m)	x 3.281	= Feet (ft)
Miles	x 1.609	= Kilometres (km)	x 0.621	= Miles

Volume (capacity)

Cubic inches (cu in; in³)	x 16.387	= Cubic centimetres (cc; cm³)	x 0.061	= Cubic inches (cu in; in³)
Pints (pt)	x 0.568	= Litres (l)	x 1.76	= Pints (pt)
Quarts (qt)	x 1.137	= Litres (l)	x 0.88	= Quarts (qt)
Gallons (gal)	x 4.546	= Litres (l)	x 0.22	= Gallons (gal)

Mass (weight)

Ounces (oz)	x 28.35	= Grams (g)	x 0.035	= Ounces (oz)
Pounds (lb)	x 0.454	= Kilograms (kg)	x 2.205	= Pounds (lb)

Force

Pounds-force (lbf; lb)	x 4.448	= Newtons (N)	x 0.225	= Pounds-force (lbf; lb)
Newtons (N)	x 0.1	= Kilograms-force (kgf; kg)	x 9.81	= Newtons (N)

Pressure

Pounds-force per square inch (psi; lbf/in²; lb/in²)	x 0.070	= Kilograms-force per square centimetre (kgf/cm²; kg/cm²)	x 14.223	= Pounds-force per square inch (psi; lbf/in²; lb/in²)
Pounds-force per square inch (psi; lbf/in²; lb/in²)	x 0.068	= Atmospheres (atm)	x 14.696	= Pounds-force per square inch (psi; lbf/in²; lb/in²)
Pounds-force per square inch (psi; lbf/in²; lb/in²)	x 0.069	= Bars	x 14.5	= Pounds-force per square inch (psi; lbf/in²; lb/in²)

CONVERSION FACTORS

Torque (moment of force)

Pounds-force feet (lbf ft; lb ft)	x 0.138	= Kilograms-force metres (kgf m; kg m)	x 7.233	= Pounds-force feet (lbf ft; lb ft)
Pounds-force feet (lbf ft; lb ft)	x 1.356	= Newton metres (Nm)	x 0.738	= Pounds-force feet (lbf ft; lb ft)
Newton metres (Nm)	x 0.102	= Kilograms-force metres (kgf m; kg m)	x 9.804	= Newton metres (Nm)

Power

Horsepower (hp)	x 0.745	= Kilowatts (kW)	x 1.3	= Horsepower (hp)

Velocity (speed)

Miles per hour (miles/hr; mph)	x 1.609	= Kilometres per hour (km/h; kph)	x 0.621	= Miles per hour (miles/hr; mph)

Fuel consumption*

Miles per gallon (mpg)	x 0.354	= Kilometres per litre (km/l)	x 2.825	= Miles per gallon (mpg)

Temperature

Degrees Fahrenheit = (°C x 1.8) + 32

Degrees Celsius (Degrees Centigrade; °C) = (°F − 32) x 0.56

* Note: It is common practice to convert from miles per gallon (mpg) to litres/100 kilometres (l/100 km), where mpg x l/100 km = 282

DISTANCE TABLES

MILES

	LONDON	Aberdeen	Aberystwyth	Birmingham	Bournemouth	Brighton	Bristol	Cambridge	Cardiff	Carlisle	Chester	Derby	Dover	Edinburgh	Exeter	Fishguard	Fort William	Glasgow	Harwich	Holyhead
LONDON		806	341	179	169	89	183	84	249	484	298	198	117	608	275	422	818	634	114	420
Aberdeen	501		724	658	903	895	789	742	792	349	571	634	924	198	911	795	254	241	848	692
Aberystwyth	212	450		193	328	425	206	354	177	375	153	222	459	526	325	90	708	525	460	182
Birmingham	111	409	120		237	267	132	161	163	309	117	63	296	460	259	283	642	459	267	246
Bournemouth	105	561	204	147		148	124	251	195	554	351	306	278	705	132	375	887	703	283	463
Brighton	55	556	264	166	92		217	180	288	573	386	286	130	697	275	468	906	723	203	509
Bristol	114	490	128	82	77	135		233	71	439	237	195	314	591	121	251	722	589	314	340
Cambridge	52	461	220	100	156	112	145		290	417	269	155	204	544	351	436	750	566	106	391
Cardiff	155	492	110	101	121	179	44	180		443	237	225	385	587	192	180	776	608	359	341
Carlisle	301	217	233	192	344	356	273	259	275		222	304	602	151	571	444	333	150	523	341
Chester	185	355	95	73	218	240	147	167	147	138		114	415	373	357	243	555	372	375	137
Derby	123	394	138	39	190	178	121	96	140	189	71		315	436	315	312	637	454	261	251
Dover	73	574	285	184	173	81	195	127	239	374	258	196		726	391	539	935	752	203	538
Edinburgh	378	123	327	286	438	433	367	338	365	94	232	271	451		723	616	233	72	650	489
Exeter	171	566	202	161	82	171	75	218	119	355	222	196	243	449		372	904	721	389	460
Fishguard	262	494	56	176	233	291	156	271	112	276	151	194	335	383	231		777	594	533	272
Fort William	508	158	440	399	551	563	480	466	482	207	345	396	581	145	562	483		183	856	671
Glasgow	394	150	326	285	437	449	366	352	378	93	231	282	467	45	448	369	114		673	488
Harwich	71	527	286	166	176	126	195	66	223	325	233	162	126	404	242	331	532	418		497
Holyhead	261	430	113	153	288	316	211	243	212	212	85	156	334	304	286	169	417	303	309	
Hull	168	348	229	143	255	223	226	124	244	155	132	98	251	225	301	285	362	250	181	217
Inverness	537	105	482	445	594	590	528	497	524	253	387	430	601	159	603	529	66	176	563	463
Leeds	190	322	173	111	261	251	206	148	212	115	78	74	269	199	268	229	322	210	214	163
Leicester	98	417	151	39	166	153	112	68	139	208	94	28	171	294	187	207	415	301	134	183
Lincoln	136	382	186	85	217	191	163	86	190	180	123	51	213	259	238	245	387	273	152	208
Liverpool	205	334	104	94	235	260	164	184	164	116	17	88	278	210	239	160	323	211	250	94
Manchester	192	332	133	81	228	247	167	154	172	115	38	58	265	209	242	189	322	210	220	123
Newcastle	271	230	270	209	348	326	295	231	310	58	175	164	344	107	370	326	257	143	297	260
Norwich	107	487	267	155	212	162	221	60	237	287	210	139	169	364	277	331	494	380	63	299
Nottingham	123	388	154	50	189	178	132	84	152	185	87	16	196	265	207	210	392	278	150	172
Oxford	56	464	157	62	92	99	66	79	107	256	128	98	129	341	139	207	461	340	134	202
Penzance	282	683	313	268	193	269	186	329	230	466	333	307	354	560	111	342	673	559	353	397
Plymouth	213	614	244	199	124	213	117	260	161	397	264	238	285	491	42	273	604	490	284	328
Preston	216	302	144	105	256	271	185	198	190	85	49	100	289	179	266	200	292	180	264	127
Sheffield	162	358	158	75	227	217	163	124	176	145	78	36	235	235	248	214	352	240	190	167
Southampton	79	530	202	128	32	60	75	136	119	322	194	164	149	416	111	231	520	415	150	281
Stranraer	410	235	342	301	453	465	382	368	384	109	247	298	483	133	464	385	199	85	434	321
Swansea	196	506	76	126	162	220	85	216	41	289	151	165	280	383	160	71	496	382	264	187
Worcester	114	428	98	26	131	168	60	116	75	211	88	65	187	305	135	155	418	304	182	152
York	196	309	197	134	273	251	215	157	242	116	102	89	269	186	290	253	323	211	223	187

VAUXHALL NOVA

DISTANCE TABLES 147

KILOMETRES

Hull	Inverness	Leeds	Leicester	Lincoln	Liverpool	Manchester	Newcastle	Norwich	Nottingham	Oxford	Penzance	Plymouth	Preston	Sheffield	Southampton	Stranraer	Swansea	Worcester	York	
270	864	306	158	219	330	309	436	172	198	90	454	343	348	261	127	660	315	183	315	LONDON
560	169	518	671	615	538	534	370	784	624	747	1099	988	486	576	853	378	814	689	497	Aberdeen
369	776	278	243	304	167	214	435	430	248	253	504	393	232	254	325	550	122	158	317	Aberystwyth
230	716	179	63	137	151	130	336	249	80	100	431	320	169	121	206	484	203	42	216	Birmingham
410	956	420	267	349	378	367	560	341	304	148	311	200	412	365	51	729	261	211	439	Bournemouth
359	950	404	246	307	418	398	525	261	286	159	433	343	436	349	97	748	354	270	404	Brighton
364	850	332	180	262	264	269	475	356	212	106	299	188	298	262	121	615	137	97	346	Bristol
200	800	238	109	138	296	248	372	97	135	127	529	418	319	200	219	592	348	187	253	Cambridge
393	843	341	224	306	264	277	499	381	245	172	370	259	306	283	192	618	66	121	389	Cardiff
249	407	185	335	290	187	185	93	462	298	412	750	639	137	233	518	175	465	340	187	Carlisle
212	623	126	151	198	27	61	282	338	140	206	536	425	79	126	312	398	243	142	164	Chester
158	692	119	45	82	142	93	264	224	26	158	494	383	161	58	264	480	266	105	143	Derby
404	967	433	275	343	447	426	554	272	315	208	570	459	465	378	240	777	451	301	433	Dover
362	256	320	473	417	338	336	172	586	426	549	901	790	288	378	669	214	616	491	299	Edinburgh
484	970	431	301	383	385	389	595	446	333	224	179	68	428	399	179	747	257	217	467	Exeter
459	851	369	333	394	257	304	525	533	338	333	550	439	322	344	372	620	114	249	407	Fishguard
583	106	518	668	623	520	518	414	795	631	742	1083	972	470	566	837	320	798	673	520	Fort William
402	283	338	484	439	340	338	230	612	447	547	900	789	290	386	668	137	615	489	340	Glasgow
291	906	344	216	245	402	354	478	101	241	216	568	457	425	306	241	698	429	293	359	Harwich
349	745	262	295	335	151	198	418	481	277	325	639	528	201	262	455	517	304	245	301	Holyhead
	615	90	145	61	196	151	190	240	148	262	663	552	180	103	412	425	433	272	63	Hull
382		579	729	673	605	604	428	842	682	816	1159	1049	555	634	925	417	872	747	552	Inverness
56	358		158	108	119	64	148	277	116	272	613	502	90	58	378	360	369	220	39	Leeds
90	453	98		82	187	138	301	187	42	119	480	369	206	105	225	510	282	109	174	Leicester
38	418	67	51		192	137	245	169	56	201	562	451	185	72	307	465	340	192	124	Lincoln
122	376	74	116	119		55	253	360	167	251	563	452	50	119	357	362	270	169	158	Liverpool
94	375	40	86	85	34		212	306	119	230	568	457	48	64	325	360	295	164	103	Manchester
118	266	92	187	152	157	132		414	254	412	774	663	203	206	518	262	517	378	127	Newcastle
149	523	174	116	105	224	190	257		198	224	626	515	354	241	299	637	439	283	293	Norwich
92	424	72	26	35	104	74	158	123		156	512	401	167	64	262	473	283	122	134	Nottingham
163	507	169	74	125	156	143	256	139	97		402	291	272	217	106	587	230	95	291	Oxford
412	720	381	298	349	350	353	481	389	318	250		126	597	560	357	925	436	396	645	Penzance
343	652	312	229	280	281	284	412	320	249	181	78		486	447	246	814	325	285	534	Plymouth
112	345	56	128	115	31	30	126	220	104	169	371	302		113	375	312	322	203	126	Preston
64	394	36	65	45	74	40	128	150	40	135	348	278	70		323	490	336	163	90	Sheffield
256	575	235	140	191	222	202	322	186	163	66	222	153	233	201		679	257	193	398	Southampton
264	259	224	317	289	225	224	163	396	294	365	575	506	194	254	422		641	515	362	Stranraer
269	542	229	175	211	168	183	321	273	176	143	271	202	200	209	160	398		161	418	Swansea
169	464	137	68	119	105	102	235	176	76	59	246	177	126	101	120	320	100		257	Worcester
39	343	24	108	77	98	64	79	182	83	181	401	332	78	56	247	225	260	160		York

VAUXHALL NOVA

INDEX

A

About this handbook 5
Accidents 35
    Dealing with 35
    First aid 36
    Recording details of 38
    Requirements of the law 38
Air cleaner element, renewal 100
Air cleaner setting 107
Alcohol and driving 56
Alternator drivebelt 98
Asbestos, precautions concerning 87

B

Bad weather (driving) 52
Battery electrolyte level 78
Battery precautions 87
Battery, starting when flat 47
Battery terminals, cleaning 107
Bodywork, checks and maintenance ... 108, 111
Bonnet release 27
Boot light 118
Brake fluid level 76
Braking system checks 100
Breakdowns 43
Breakdowns, emergency kit 49
Broken windscreen, dealing with 42
Bulbs, renewal 113
    Boot light 118
    Courtesy light 117
    Direction indicator repeater light 115
    Direction indicator (front) 114
    Engine compartment light 118
    Foglight (front) 115
    Headlight 114
    Luggage compartment light 118
    Number plate light (rear) 117
    Rear lights 116
    Sidelight (front) 114
Buying and selling 9

C

Caravan, towing 54
Central door locking system 34
Changing a wheel 44
Child safety 59
Cigarette lighter 27
Clock 19
Clutch checks 99
Clutch pedal adjustment 103
Contact breaker points 96
Controls and equipment 15
Conversion factors 144
Coolant antifreeze concentration 94
Coolant draining, flushing and refilling 105
Coolant level 75
Coolant, renew 105
Courtesy light 117
Crime prevention 67

D

Dimensions and weights 12
Direction indicator (front) 114
Direction indicator repeater light 115
Distance tables 146
Distributor, clean 97
Door locks 34
Door mirror switch 26
Driveshaft checks 99
Driving:
    Abroad 61
    In bad weather 52
    On motorways 53
    Safety 51

VAUXHALL NOVA

INDEX

E

Economical driving 65
Electric door mirror switch 26
Electric windows......................................27
Electricity (mains) and electrical equipment,
precautions concerning 88
Emergencies...35
Engine compartment 93
Engine oil and filter, renewal 93
Engine oil, disposal of 88
Equipment..15
Exhaust system 95

F

Fanbelt 98
Fault finding 123
    Brakes..126
    Electrics..128
    Engine 123
    Gearbox, transmission and clutch 125
    Suspension and steering......................127
Faults while driving 99
Fire hazards, precautions concerning 87
Fire in vehicle, dealing with.....................42
First aid 36
Foglight (front) 115
Foglight switch 26
Fuel gauge 18
Fuel pump filter screen 95
Fumes, precautions.................................87
Fuses 118

G

Gearbox (manual)...................................20
Gearbox checks 99
Gearbox oil..104
Glossary 129

H

Hazard flasher switch 25
Head restraints 31
Headlight 114
Heated front seats switch 26
Heated rear window switch 26
Heating 23
Hinges 98
Horn 15, 16, 79

I

Ignition HT voltage, precautions concerning..88
Ignition switch/steering lock....................20
Instruments 18, 99
Interior 111
Interior mirror 29

J

Jacking and vehicle support 88
Jacking and wheel changing 44
Jargon 129
Jump lead starting 48

L

Leaks (fluid), checking for 80
Lights, checking operation 79
Lights switch (exterior lights)...................21
Local radio frequencies 139
Locks 34
Loudspeaker balance control 26
Lubricants and fluids...............................69
Luggage compartment light 118

VAUXHALL NOVA

INDEX

M

Mains electricity, precautions 88
Maintenance 81
Maintenance schedule 82
MOT, preparing for 121
    Brakes 121
    General 122
    Lights 121
    Seatbelts 122
    Steering and suspension 121
    Tyres and wheels 122
Motorway breakdowns 44
Motorway driving 53
Multi-function switch 20, 22

N

Number plate light (rear) 117

O

Oil and filter, renewal 93
Oil level 74
Oil pressure gauge 19

P

Parcel shelf 33
Plugs (spark) 70, 101
Punctures 44
Push starting 48

R

Radio frequencies 139
Rear foglight switch 25
Rear lights 116
Rear parcel shelf 33
Rear seats 32
Rear view mirror 29
Relays 119
Rev. counter 18
Road test 99
Routine maintenance 81

S

Safety (child) 59
Safety (driving) 51
Safety first! 86
Scratches 112
Seat adjustment 31
Seat belts 29
Security 67
Selling 11
Service schedule 82
Service specifications 69
Service tasks 93
    250 miles (400 km)/weekly 93
    9000 miles (15000 km)/12 months 93
    18000 miles (30000 km)/2 years 100
    2 years, regardless of mileage 105
    36000 miles (60000 km)/4 years 106
    Seasonal servicing 107
Servicing 81
Servicing notes 152
Sidelight (front) 114
Skid control 56
Spare parts 89
Spark plug, renewal 101
Spark plug specifications 70
Specifications 69
Speedometer 18
Starting a car with a flat battery 47
Steering checks 99

VAUXHALL NOVA

INDEX 151

Steering lock 20
Sunroof 28
Suspension checks 99
Switches:
    Electric door mirror 26
    Front foglight 26
    Hazard flasher (warning) 25
    Headlamp range 22
    Heated front seats 26
    Heated rear window 26
    Horn 15, 16
    Ignition 20
    Indicators 20
    Lights 21
    Loudspeaker fader 26
    Multi-function 20, 22
    Rear foglight 25
    Steering column 20, 22
    Tailgate wash/wipe 23
    Windscreen wash/wipe 22

T

Tachometer 18
Tailgate wash/wipe switch 23
Temperature gauge 18
The Nova family 7
Throttle/choke linkage 95
Tools 109
Tow starting 48
Towing 46
Towing 54
Trailer, towing 54
Troubleshooting .... see Fault finding
Tyre pressures 71
Tyres, checking 76

U

Underside 99, 108

V

Vandalism, dealing with 42
Ventilation 23
Voltmeter 19

W

Warning lights 19
    Brake fluid, low level 19
    Choke 19
    Direction indicator 19
    Engine control indicator 20
    Handbrake 19
    Headlight main beam 19
    Ignition 19
    Oil pressure 19
    Trailer indicator 19
Wash/wipe switch 22
Washer fluid level 78
Weights 13
Wheel changing 44
Windscreen breakage, dealing with 42
Windscreen wash/wipe switch 22
Wipers and washers, checking 79
Women drivers, advice to 57

VAUXHALL NOVA

SERVICING NOTES

Date	Action

VAUXHALL NOVA

SERVICING NOTES

Date	Action

VAUXHALL NOVA

154 SERVICING NOTES

Date	Action

VAUXHALL NOVA

SERVICING NOTES 155

Date	Action

VAUXHALL NOVA

SERVICING NOTES

Date	Action

VAUXHALL NOVA

SERVICING NOTES

Date	Action

VAUXHALL NOVA

SERVICING NOTES

Date	Action

VAUXHALL NOVA

SERVICING NOTES

Date	Action

VAUXHALL NOVA

160 OWNERS WORKSHOP MANUALS

Nova Manual

Like this handbook, the above manual has been written specifically for your Nova. Most models are covered from 1983 to February 1992, the only exceptions being the Van, those fitted with Diesel engines, and the revised Nova range introduced in February 1992.

The manuals are written in the same easy-to-follow manner as the handbooks and enable you to tackle even major overhauls and repairs yourself. They will advise you on which jobs are suitable for the DIY mechanic and often suggest practical alternatives to the specialist tools that are so often recommended.

With the ever increasing cost of garage labour, think of the money you could save!

Haynes Manuals are available from most good motor accessory stores and bookshops, or direct from the publisher.

 **HAYNES**

Haynes Publishing, Sparkford, Nr Yeovil, Somerset BA22 7JJ Telephone 0963 440635

VAUXHALL NOVA